BIOLOGY & MANKIND

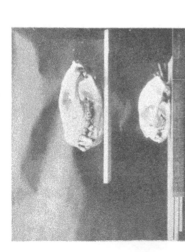

I.

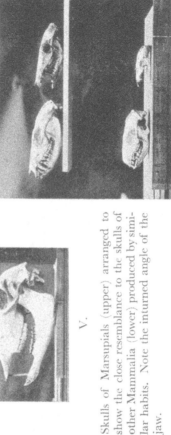

II.

III.

IV.

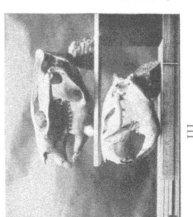

V.

Fig. 1.

Skulls of Marsupials (upper) arranged to show the close resemblance to the skulls of other Mammalia (lower) produced by similar habits. Note the inturned angle of the jaw.

I. Cat type, Tasmanian Devil (*Sarcophilus*) and Cat (*Felis*). II. Dog type, Tasmanian Wolf (*Thylacinus*) and Dog (*Canis*). III. Rodent type, Wombat (*Phascolomys*) and Beaver (*Castor*). IV. Insectivorous type, Opossum (*Didelphys*) and Hedgehog (*Erinaceus*); Bandicoot (*Perameles*) and Mole (*Talpa*). V. Herbivorous type, Kangaroo (*Macropus*) and Sheep (*Ovis*).

BIOLOGY & MANKIND

BY

S. A. McDOWALL, B.D.

Chaplain and Senior Science Master
at Winchester College

CAMBRIDGE
AT THE UNIVERSITY PRESS

1931

CAMBRIDGE UNIVERSITY PRESS
Cambridge, New York, Melbourne, Madrid, Cape Town,
Singapore, São Paulo, Delhi, Tokyo, Mexico City

Cambridge University Press
The Edinburgh Building, Cambridge CB2 8RU, UK

Published in the United States of America by Cambridge University Press, New York

www.cambridge.org
Information on this title: www.cambridge.org/9781107605039

© Cambridge University Press 1931

First published 1931
First paperback edition 2011

A catalogue record for this publication is available from the British Library

ISBN 978-1-107-60503-9 Paperback

CONTENTS

LIST OF ILLUSTRATIONS

PREFACE

A Preface may be explanatory; it may be apologetic; it may be frankly propagandist. This book is intended to give a short but fairly complete summary of the development and present position of the studies of Evolution and Heredity, suitable for the general reader as well as for the biological student who needs a foundation on which to build; but it also represents one portion of an educational experiment which I dare to believe important. Since it has been tried with marked success for more than a dozen years, I intend to use the opportunity of a preface to give a brief explanation of that experiment in a frankly propagandist spirit; and further to urge that a knowledge of the facts of heredity and their bearing upon social and political life is essential to every man who cares about the future of England.

I am firmly convinced that the biological specialist must start with a thorough training in chemistry and physics; I am equally convinced that a general introduction to the facts and theories of biology ought to be the crown of a general science course for boys whose chief studies are classical or historical. In both cases (and ideally in the case of chemists, physicists, and mathematicians also) the later this general biology can come the better. The living organism is the meeting-place of physical and chemical phenomena, yet from the living organism spring history, art, science, religion and language. In the presence of life the water-tight compartments of our teaching are broken down.

With these convictions my colleagues and I have built up for the classical boys a general science course occupying two to three periods a week for five years, comprising Physical Geography, Hydrostatics and Heat, Chemistry, Physics, Geology and Astronomy, and General Biology, of which the Physical Geography and Geology are alternatives: a boy who starts with the Physical Geography omits the Geology and Astronomy. In the earlier stages there is regular laboratory work, but the large size of the General Biology class makes laboratory work impossible, and we have to be content with the epidiascope, lantern-slides, projected microscope-slides and demonstrations. It is with this last-year course that we are now concerned.

The first term is spent in a general study of the conditions of "knowing" through a survey of the development of Greek philosophy, designed to show the gradual separation of the spheres of Natural Science, Philosophy, and Religion and to indicate both the scope and interrelations of these, and the general problem of Reality. Then the development of Science is briefly traced, with constant references to History, through the Hellenistic and Arab cultures and the Renaissance down to the present day. A few lives—Roger Bacon, Newton, Faraday, Pasteur—supply a biographical element.

The second term begins with a study of the living organism as a physico-chemical phenomenon. The boys are already familiar with the terminology and concepts of physics and chemistry, and no difficulty is experienced on this account.

Amoeba and the *diatoms* furnish the text for a general examination of the sources of vital energy; *paramoecium* and *vorticella* introduce conjugation and tropisms; *hydra*

and *obelia* sexual reproduction, alternation of generations, physiological division of labour, and the detail
of nuclear division. The frog gives a general acquaintance
with vertebrate structure, and its embryology supplies
an illustration of von Baer's Law of Recapitulation.
Vertebrate tissues are next examined: connective tissues
(from blood to bone), glandular tissues, muscular tissues,
nervous tissues. This leads on to the rudiments of human
physiology, including digestion, excretion, and respira-
tion. The ductless glands next serve as an important
link between physiological and psychological processes.
A short study of psychology concludes the term. Of
necessity this is restricted to a few relevant topics:
Locke's transfer of the Platonic Idea from Heaven to
the human mind, and the resulting development of
Idealist philosophy; the attempts to explain all mental
activities in terms of the Reflex and Conditioned Reflex; tropistic theories; epiphenomenalism and psycho-
physical parallelism; and at the end a couple of lectures
on the root-conceptions of the old and the new psychology.

Twice in the term, however, we make brief use of our
previous study of philosophy. Our physico-chemical
enquiries land us in completely mechanistic conceptions,
and these we consider from a philosophical stand-
point; our examination of the new psychology lands us
in a more subtle determinism, and we discuss the reason
why this becomes self-destructive when pressed to its
logical issue.

The lectures of the third term are nearly, but not
quite, represented by the material of this book.

Not quite, for the course ends with another examination into the philosophical and religious aspect of our

work. A third time we have been landed in determinism, this time as the result of our study of evolution and heredity. An elementary enquiry into the cause of the fact that physiology, psychology, and heredity each in turn lands us in a rigid determinism reveals the existence of two possible explanations. The first is, that the result is due to the real nature of the universe, the second that it is due to the method of abstraction which science adopts in studying the universe. This dilemma leads us back to a general survey of the scientific method, and forward to a slight examination of the phenomenon we call personality; and finally we attempt to construct an outline philosophy of biology which shall neglect *no* phenomena—not even those of the mind and spirit.

It was my original intention to publish the whole of the biological portion of this course in a single volume. For various reasons it was decided not to do this, unless the future should disclose a demand. The book would have been bulky; the earlier material could for the most part be readily obtained elsewhere; only the mode of presentation was of special interest. Finally, it would be quite possible to issue the whole text later, if it seemed desirable.

For these reasons it was decided to publish only the portion dealing with evolution, heredity, and the biological aspects of social and political conditions; at all events for the present. The final section of this dealing with the philosophical problem, when divorced from the previous surveys of the same theme, seemed rather in the air, and I decided very reluctantly to omit it. Thus the present volume covers compactly the material needed by the serious student of biology who wishes to survey the development of our theories of evolution and heredity (I make all my biology specialists attend this portion of the course) together with a discussion of the interplay between biological laws and the sicknesses of civilised society.

But I would urge, with all the force at my command, that such matters do not fall within the province of the biologist alone. The problems and the disasters of to-day, social, political, economic, are in the long run biological problems. Because we have been governed by politicians who know nothing of biology; because we ordinary citizens are content to remain in such abysmal ignorance of the inevitable effects of mistaken ideas and mistaken legislation that we allow our leaders of all parties to enact social laws which go clean in the face of inescapable biological laws, and to shirk the responsibility of taking action when a clear method of escape lies open; because, blind ourselves, we trust blind guides; because we cannot think further ahead than the life of a parliament while Nature works by generations and tens of generations; we are face to face with disaster.

It is the business of every citizen to know the established facts of heredity. Being absolutely convinced of this, I have for a dozen years taken care that the boys who reach the top of the school which I have the honour to serve shall know those facts, and shall learn to realise that physics and chemistry and biology are not only matters for the expert, but inextricably interwoven with every activity of human life. It is a small beginning; but there is no question that the average boy of seventeen or eighteen can and does understand the argument, is intensely interested, and finds that he has come to look on the world from a new point of view. In the hope that others will try the experiment with equal success—for any biologist can easily give such a course of lectures if he will take the trouble to collect the material—I publish, in what I hope is a convenient form, some of the data which I have collected from various sources. In the same hope I write this preface: I beg my fellow-teachers to believe that, not egotism, but the necessity of avoiding circumlocution, is responsible for the deplorable number of "I's" in it.

Naturally such a course changes, and I hope improves, every year: this is necessary if it is to remain alive. But although experience still suggests improvements of detail I have found that I have made few large alterations in the last few years; so that the form seemed sufficiently defined to justify publication.

It is perhaps worth while to add that I always caution my audience against accepting my views, not once, but over and over again; and I try to take particular care that the transition from fact to deduction and theorising is observed. I explain that, subject to the errors of human frailty, the facts I give them are correct; and I explain that I am personally driven to draw certain conclusions from these facts, but that I may perfectly well be mistaken; that the last thing I want them to do is to accept my views, but that I do want them to weigh the facts and think for themselves. If I did not say what I thought myself quite honestly, the lectures would be ineffective. It is better to be rash than woolly.

One slight disadvantage involved in the publication of only the latter part of the course lies in the use of terms which were explained in the earlier stages of the work. Many of these are made clear as they occur; but in order to avoid the interruption of the argument by cumbrous explanations, a glossary of such terms has been inserted at the end.

It is impossible fully to acknowledge my indebtedness to this book and that. In lectures so eclectic the range of borrowing must be wide; and many of the facts are so deep-sunk in my mind that I have not the least idea of where I first got them. But I have gained so much from East and Jones' *Inbreeding and Outbreeding*, from East's *Heredity and Human Affairs*, from Whetham's *The Family and the Nation*, and from J. Arthur Thomson's *Concerning Evolution*, that special mention must be made of these. Punnett's *Mendelism* has been an unfailing stand-by; most of all I have drawn on the inspiration of the teaching of William Bateson, and the friendship of W. H. R. Rivers, Hugh Wingfield, and William Brown. I can only advise my readers to consult the works I have mentioned, and others such as Ruggles Gates' *Heredity in Man*, and R. A. Fisher's recent volume, *The Genetical Theory of Natural Selection*, where they will find a wealth of material that will open their eyes both to the extent of the recent advances in our knowledge, and the vast importance of the issues raised.

The illustrations are chosen in order that the reader may be able to grasp the appearance of the organisms under discussion. I wish to express my gratitude to Messrs George Allen and Unwin, Blackie and Son, Longmans, Green, and Macmillan, and to the Clarendon Press and the Macmillan Company of New York for permission to include figures from works published by them.

In conclusion I would say one word to the general reader, for whom this volume is designed at least as much for the teacher and the student of biology. Although I believe with all my heart that it is a crying shame that boys and girls should leave school with no idea of the interesting and important knowledge that has been gained about Man, his make-up and his environment, I believe equally that the ignorance of the intelligent adult citizen concerning these things is a national disaster. In publishing this book I dare to hope that some men and women who care for England may read it, and may find therein the material of a new inspiration. Though I utterly disbelieve in the present phase of Democracy, I am convinced that out of it something really great will yet emerge, if we will but take thought in time.

Biological laws are always at work, even though we neglect to study them. Shutting our eyes will not stop their passionless, stern course.

The greatest need of England, and of England's politicians, is a knowledge of the biological factors upon which legislation may impinge; and it is mainly the younger generation that must acquire and demand such knowledge.

However inadequately, the following pages represent an attempt to meet that need.

S. A. McDOWALL

WINCHESTER 1931

Chapter I

EVOLUTION

There are only two possible ways of accounting for the present state of affairs in our world. Either the processes by which it was achieved were continuous, or else they were discontinuous. The method of science is entirely dependent on the assumption of continuity. One of the most striking changes that has come about in modern civilisation is this attitude of mind which tries to trace back language, art, religion, and every form of human activity to its primitive manifestations. The scientific method has revolutionised our thought by imposing upon us the expectation of continuity. So long as philosophy is brought in to supply the required corrective by reminding us that continuity is only continuity of process, and that this process itself requires an explanation, there is great gain; but it must always be remembered that a process is not self-explanatory, nor can its real meaning be found in its beginnings. The meaning of a process must be sought in the highest terms of the series, not in the lowest; the business of science is merely to trace it backwards to its first manifestations, and to establish the continuity of the laws which govern it. Uniformity is at once the postulate and the goal of science.

We therefore naturally tend to prefer the idea of a continuity which covers the development of the living organisms as well as of the stars and planets and elements; and, further, a continuity which leaves no gap between the living and the non-living.

The conception of continuity must take one of two forms; the conception of discontinuity has also taken two forms, the first very naïve, the second rather a washy compromise.

Continuity. We may imagine, as Arrhenius has imagined, that when the earth reached a suitable condition, microscopic spores, drifting everywhere in space under the pressure of light, and preserved by low temperature and lack of oxygen, germinated as they fell on the earth, and introduced life into the planet. This may or may not be true: even if it is true, it only shifts the question one stage farther back, for we at once want to know how and where these spores first originated.

The other possibility is that under certain peculiar conditions a colloid aggregation of molecules (a micella), in a strangely unstable and labile condition, comes into being, bridging the gap between the living and the non-living. Professor Baly's recent success in synthesising formaldehyde and sugars, and in combining nitrites with formaldehyde to make organic compounds of nitrogen, by the use of short wave-lengths of light and catalytic agents, has gone a long way towards bridging this gap. Though it remains true that life as we know it can only be derived from life, we seem to be moving in the direction of a possible synthesis of living matter. The idea that we shall ever be able to create a living animal is almost fantastic, and there is no danger of our making a Frankenstein monster; but it is not beyond the bounds of possibility that we may one day be able to synthesise a kind of protoplasm in the laboratory: a substance more or less living, which will behave more or less as an organism. This must be the hope of the biologist, and the goal of biophysics and biochemistry.

Such a goal, whether reached or not, implies a thoroughgoing belief in continuity. The hope does not in the least imply an expectation that we shall eventually be able to explain mental and spiritual things in terms of matter: to think so is to be guilty of the philosophical error of explaining the highest in terms of the lowest, instead of the lowest in terms of the highest. But it does, I think, impress the need of one single ultimate explanation; and if you may not explain mind or spirit in terms of matter, you must needs explain matter in terms of mind and spirit.

Discontinuity. The older, naïve form of this explanation was that the Almighty created each species. The newer, due mainly to a desire to preserve a conception erroneously supposed to be religious, imagines one great act of creation of LIFE intervening between the dead matter of the stellar and planetary systems already created, and the world of living things as we know it, which evolved from the primitive life thus created. This is a compromise which satisfies neither the scientific mind nor the primitive concepts of God which characterise the infancy of religion; yet we cling to it because here, as elsewhere, we are very loth to put away childish things.

It is perhaps worth observing that ideas which suggest continuity and unity in the Deity are likely to be far truer and more worthy than ideas which imply spasmodic activity and effortful contrivance at each stage.

Since the older idea was that God created species, and since Darwin called his epoch-making book *The Origin of Species*, it is well that we should try to define species.

Unfortunately they almost defy definition. It is no help to say that species were what God originally created, for we were not there to see. A once-prevalent view was based on the physiological fact that species can interbreed and produce fertile offspring. The fact is true; but it is also true that fertile hybrids between different species exist, while sterility between individual members of the same species is not uncommon. Again, though most species are perfectly well defined, in many cases species shade off into one another, more especially among plants, so that though the typical members are easy to detect, it is not always easy to say to which of two species a certain individual may belong. Also, a species may be dimorphic, having two norms.

Linnaeus, the father of modern species, was uncomfortable about the whole situation. He cut the Gordian knot, however, with a new version of the legal maxim "de minimis non curat lex"; telling his followers that "varietates levissimas non curat botanicus".

Actually, a species is an entity created by the biologist for his own convenience. In most cases it does correspond with the facts; in some the species is not yet really established, but is very variable; and in others it appears to be splitting up. Broadly, however, a species consists of a group of organisms normally breeding together, which only vary from the type within limits, and of which the varieties are connected to the norm by a more or less continuous series.

Deciding then in favour of a wholehearted doctrine of continuity, as the more scientific and more probable explanation, we may first note briefly some of the lines of evidence which offer direct proof that evolution has actually occurred in the organic world, and then turn to the theories which have been suggested in explanation of the fact.

It is of the first importance to realise that the facts are independent of the theories which attempt to account for them. It is the commonest thing in the world to hear someone say, "Oh, Darwin was all wrong in his theories [a gross exaggeration, by the way] and therefore evolution is an exploded idea". Such people merely expose their own inability to think clearly. The facts are there: the theories which have been put forward to account for them are various, and none has yet proved completely adequate: perhaps there is a measure of truth in all of them and their error lies in exclusiveness; perhaps not. At any rate, there have been many theories, and as knowledge increases there will be many more. What is important is, that we should not confuse the facts with the theories put forward to explain them.

Of the fact of evolution there is no possible shadow of doubt.

Since any theory of evolution must find its ultimate foundation in the facts of heredity, we shall devote careful study to that, finding new light on the older evolutionary difficulties. This will inevitably bring us back to the details of nuclear division. After that we can study heredity in man, and then, and not till then, we shall be ready for the wider interest which will open up from our doctrine of continuity. Man's social customs and laws, past history, and future prospects, all have their biological aspect, and are full of interest from this point of view. The neglect to consider this biological aspect has been in the past a misfortune: now it is a crime.

An introductory study, however slight, is better than none at all, provided that the laws of heredity are first thoroughly grasped. Whether we take account of them or not, there is no blinking the fact that they are operating all the time.

Before we enumerate the kinds of evidence that have resulted in the doctrine of organic evolution we may remind ourselves that another process of continuous change, the evolution of solar systems, of planets, and of the structure of the earth's crust, has already accustomed our minds to continuity of development; and that recent progress in physics and chemistry has again emphasised this continuity in presenting us with the electron and the quantum as the substrata of matter and energy, while wave-mechanics looks like bridging the gap between these two conceptions.

When we look at living organisms we find an extraordinarily close nexus. One or two illustrations will serve to fix the fact in our minds.

The oft-quoted case of the relation between cats and the clover crop, since the cats prey on birds and field-mice, which prey on humble-bees, which fertilise the clover, can be paralleled by innumerable instances. Professor J. Arthur Thomson mentions many in his lectures, *Concerning Evolution*, from which we may select a few.

The leaf-cutter ant lives underground on cultivated fungi. The ants take bits of leaves down, and chew them into a compost which forms mushroom beds. In those beds the fungi grow—they are known nowhere else. At the nuptial flight the queen carefully takes the fungus spawn, rolls it into a little ball, and puts it into her mouth. In the new colony she carefully sows it.

The death-watch beetle, which is responsible for the worm-holes in wooden beams, lives on the rather indigestible diet of wood. It carries a kind of yeast in two pockets in the alimentary canal which apparently does a preliminary digestion for it. No yeast is to be found in the embryo, yet when the larva hatches out there is the yeast, and wood is digested as before. Actually the female, when she lays the egg, squirts a little fluid containing yeast from two recesses beside the ovipositors, and this sticks to the rough outside of the egg. When the grub hatches, it eats its own egg-shell, and thus the yeast is introduced into the alimentary tract again.

So too, pine trees are enabled to live in an unpromising situation by the fungi which supply their roots with nitrogen, while receiving carbonaceous food from the tree. In the same way heather is able to maintain itself on moorland soil.

A parasite which kills its host is ill-adapted, but at least it usually makes provision for transfer to another host; perhaps by a mosquito which preys upon that host; perhaps by a snail upon which the desired host itself preys. The interdependence of organic lives becomes the more striking the more we study it. Instances of this kind of adaptation, taken in connection with heredity and the variability of organisms, and with the struggle for existence, led Darwin on to formulate his theory.

The biologist is not, as most people seem to think, a person who, when an obscure and almost invisible beetle is presented to him, can at once say "That is *Euplectus ambiguus*", or "Ah! *Hypocyptus longicornis*, I think". He may be a specialist in some particular branch, but essentially he is a person who looks on the whole of organic life with the idea of continuity in his mind; who has enough knowledge to be capable of weighing a large amount of evidence and drawing the correct conclusion; who knows enough chemistry and physics to be able to appreciate the needs and processes of the organism—and this demands a very thorough knowledge indeed; and who has, if possible, that last touch of genius which can see a new truth.

There is much outcry about the dearth of biologists. No doubt finance and opportunity have something to do with it, but the great biologist is born, not made; and the combination of gifts required causes him to be a very rare bird.

But any man of average intelligence can learn to appreciate the main issues of biology, and to see that in such and such a direction the neglect of biological principles will lead to national disaster. This book, indeed, represents an individual attempt to remedy the ignorance which a mistaken idea of education has imposed not only upon the classic and the historian but also upon the chemist and the physicist.

We may begin our summary of the evidences of evolution by setting out seven heads under which it may be conveniently studied—classification, morphology, palaeontology (or distribution in time), embryology, geographical distribution (or distribution in space), variation under domestication, and spontaneous mutation. Only one or two illustrative examples of each can be given, but it will be realised that the amount of evidence available is actually very large indeed.

Classification. Aristotle laid down the first serious classification of living things, for grouping is the only possible method of understanding and remembering, and it is natural to the systematic mind. Later, Linnaeus, Cuvier and others determined more adequate groupings—the former arrived at the idea of the Natural Orders of plants, though he did not fully understand them; the latter initiated a reasonable classification of animals.

The first question that we ask is, What features does the systematist select, and why? Take the porpoise and the fish, for instance. Both have the same general shape; both have tails, and paddles behind the throat; both have a great many teeth. But the porpoise brings forth its young alive, and suckles them; and it breathes air by means of lungs. The systematist says these features are much more important than mere general shape. Anything which lives in the water, completely immersed, must have a streamline shape, and the means of balancing and propulsion. Even a seal is beginning to get the stock outline. But the bones inside a whale's flipper are the bones of a man's hand; they are quite different from the primitive arrangement of a fish. Though paddles of this shape are necessary to whale and fish alike, the inside detail need not be the same. It is in fact the details which are *not* important to the mode of life of the organism, and which therefore are not likely to be adaptively changed, that will often give to the systematist the safest guide: important things, like general shape, are laid down by the necessities of living. To live in the air wings are needed; but the wings of a bat and a butterfly are radically different. The organs are analogous, but not homologous.

The persistence of unimportant detail is seen very prettily in the marsupials, a primitive order of mammals which once dominated the world, but which were exterminated or replaced by the later orders, except in Australia.[1] Here the land bridge sank before the later types got there, so the marsupials had it all their own way. They developed very much along the lines of the

[1] Two very retiring species do survive in the forests of South America also.

different orders of later mammals, becoming carni-
vorous, herbivorous and so forth; and their skulls show a
resemblance to those of the parallel types which is sometimes extraordinarily close (fig. 1). This illustrates the fact
that structure must depend on, and be adapted to, mode
of life. The Tasmanian wolf's skull is almost identical with
that of an ordinary wolf, in everything but minutiae; yet
the two are much less closely related than a whale is to
an elephant. The Tasmanian devil's skull is very like that
of an ocelot, or any other medium-sized cat. The wombat
has adopted the rodent type and its skull closely re-
sembles that of a beaver: still more closely that of the
capybara, which was so welcome to the Swiss Family of
nursery fame. The kangaroo's is a very good imitation of
a sheep's, that of an insect-eater has the general shape,
and the sharply cusped, interlocking teeth of a hedgehog
or mole. Yet practically all the marsupials have a
curious in-turning of the angle of the lower jaw found in
no other mammals, which makes it possible to distinguish these skulls at a glance, and to group them
together. Important matters have been adapted to the
mode of life: this trivial ancestral trait remains unchanged.

Just as different structures may assume the same appearance for the performance of a particular function
(analogous organs), so the same structure may take on
quite a different function in different creatures. A
modified gill-slit persists in man in the form of the
Eustachian tube which leads from the middle ear to the
throat; the lungs were once a swim-bladder; the vocal
cords are supported by the remains of gill-arches, and
so on. Such structures are *homologous*.

The wings of a bat, a bird, and a pterodactyl are

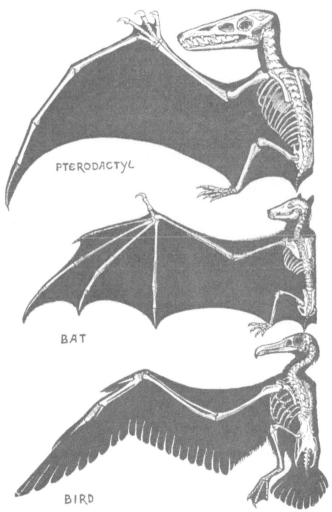

PTERODACTYL

BAT

BIRD

Fig. 2. Wing of reptile, mammal and bird.

homologous, in so far as they are all supported chiefly by the fore-limbs (though the hind-limbs are also involved, except in the bird); but *analogous* in detail, since the

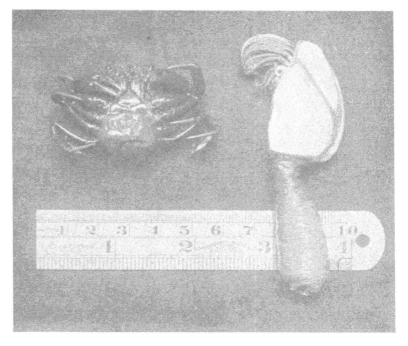

Fig. 3. Three crustacea. Shore crab with the parasitic Rhizocephalan *Sacculina* beneath the tail, and *Lepas*, the goose-barnacle. The globular sac of *Sacculina* is nourished by a system of tubes which spread over the alimentary canal of the crab. The drain on the crab's energy inhibits moulting and growth.

bat uses four fingers to stretch the membrane, the ptero-dactyl only one, while the bird has no membrane at all, its feathers springing from a structure supported by the forefinger, with additional traces of the first and third (fig. 2).

A classification which takes account of these things, instead of falsely building upon the insecure foundation of general shape, is found to take the form of a tree.

Some forms are generalised, and many of these have persisted unchanged from distant geological ages. The generalised types of one family approximate to the generalised types of another family, and the same is true of orders. As they get nearer the trunk, the main branches draw nearer to each other, and we begin to see the relationship between branch and branch. Again, though in a group like the Crustacea no one could guess the existence of a relationship between the adults of such extreme types as the goose-barnacle, the *Rhizocephala*, and the crab, yet a continuous series of intermediates links them by a chain of affinity (fig. 3).

Now a tree-like classification inevitably suggests a genealogical tree: the most obvious explanation of these relationships is through a line of descent.

Morphology. We have already noticed that the paddle of a whale has the bony structure of an ordinary hand and arm—inexplicably unless we hold a theory of descent from a land mammal. Endless similar cases might be quoted. The nautilus-shell, the cuttle-bone, and the belemnite show little resemblance; yet if we look at the nautilus we see the squid-like creature living in the last-formed chamber, while keeping touch with all the others by the "syphon", and using them as float-chambers. In *Spirula* the shell has the same structure, but is retained within the body. In the fossil *Spirulirostra* the shell is straightening, and there is a little horny projection; an arrangement recognisable in the cuttle-bone. In the belemnite the shell is straight, and the spike or rostrum is very pronounced. Such structures indicate different lines of evolution from a common type; evolution whose possibilities are further shown by the shell-less octopus, the sea-pen (the horny rostrum of *Loligo*) and the new and different shell of the paper nautilus (figs. 4 and 4*a*).

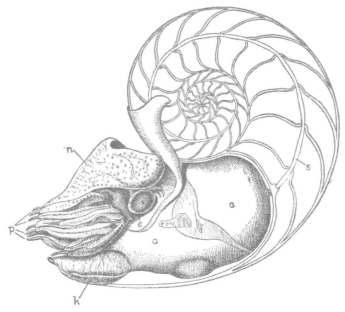

Fig. 4. Chambered nautilus, *Nautilus pompilius*, in natural position, the shell sectioned. *a*, mantle; *e*, eye; *f*, horny girdle for adhesion of mantle to shell; *k*, funnel; *n*, hood; *p*, protruded tentacles; *s*, siphuncle for communication with inner chambers of shell.

So, too, the persistence of vestiges, representing organs which once had a function but are now dwindling to nothingness, is hardly explicable on any other assumption than that of evolutionary change. Some people can still move their scalps and ears freely by means of

muscles which in most have atrophied. A few still retain
the muscles for wagging the tail or even have more tail
than the normal stump. The tiny fold in the corner of
the eye is the remnant of the third eyelid, or nictitating

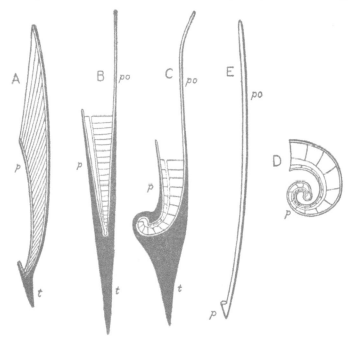

Fig. 4 a. Diagrammatic sections of dibranchiate shells. A, *Sepia*;
B, *Belemnites*; C, *Spirulirostra*; D, *Spirula*; E, *Ommastrephes*. *p*,
chambered phragmacone; *po*, proöstracum; *t*, rostrum.

membrane, which lubricates the eye of a cat and gives
an air of impassive superiority to a parrot. Near the
top of the ear, where the cartilage folds over at the edge,
can be felt a small triangular projection which was once
the point of the ear as we see it in some apes and

monkeys. Some people have extra long arms, like an ape, and always the hair on the upper arm runs downwards, while that on the forearm runs across outwards, as in the apes. This arrangement allows the rain to run off when the arms are held above the head for protection.

Infants have many ape-like characteristics, such as the inturned soles of the feet in sitting, the opposable big toe, and the power of hanging for two or three minutes by the arms when only a couple of weeks old. Instances might be multiplied endlessly: the snakes have no limb-girdles, but one, the python, has useless bones repre-

Fig. 5. *Amphioxus lanceolatus* from the left side, about twice natural size.

senting the pelvic girdle; such vestiges are found also in whales. Children are sometimes born with hare-lip. This represents the persistence of the embryonic stage when the nostrils are connected to the mouth by a groove, as they are permanently in the sharks. Cleft palate, another throwback to an earlier stage, may also be present. The appendix, though it has a new function in helping peristalsis of the intestine, was primarily designed for aiding in the digestion of cellulose, and is now hardly more than a troublesome anachronism. Thus, then, the study of structure, whether of functional organs or of functionless vestiges, strongly suggests the evolutionary explanation.

In connection with this argument and the previous one, we may note in passing the existence of a number of forms whose structure indicates that they are links between two great groups; for links are by no means always missing, even among living types. Such, for instance, is the somewhat fish-like lancelet (*Amphioxus*) (fig. 5), which is more like a worm than a fish in its blood-system,[1] and is certainly not a fish. It has the primitive skeletal structure known as a notochord, a dorsal nervous system, muscle segments, and a pharynx with very numerous gill-slits. The vertebrates must have passed through a stage much like this. Yet it shows some close affinities also with the worm-like *Balanoglossus* (fig. 6), which like it has pharyngeal gill-slits and probably a notochord; as have also the sea-squirts (fig. 7). These last begin life in a form not unlike a tadpole, but they finally stand on their heads, and suffer a sea-change into something very strange (fig. 8). Again that shy creature *Peripatus* (fig. 9) is in some ways intermediate between a worm and a caterpillar, and may link the annelids and arthropods. Here again we find a suggestion of continuity.

Palaeontology. The study of fossil forms leads to a similar conclusion.

We must remember that the record cannot possibly be complete. In the first place "There rolls the deep where grew the tree", and we cannot get at the fossils. In the areas we can explore, a few scratches in the Siwaliks, rather more in Patagonia, very little in Siberia or central Africa or the Middle East or China are, to say the least, inadequate, even when backed up by the knowledge derived from a more thorough investigation of western Europe and North America. But this is not the chief trouble. The likelihood of an animal getting

---

[1] This resemblance is probably more superficial than has sometimes been imagined.

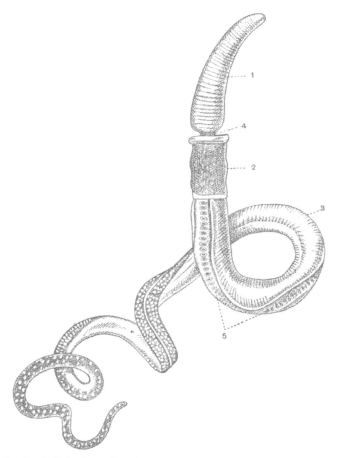

Fig. 6. *Dolichoglossus kowalevskii* × 1. (A Balanoglossus-like Hemichordate.) 1, proboscis; 2, collar; 3, trunk; 4, mouth; 5, gill-slits.

fossilised is small. It will have to die and fall on a medium that is not too soft and not too hard to take a cast—that is to say, the medium must be soft, yet harden sufficiently before decay goes too far. It must be subjected to no very great disturbance afterwards such as would disintegrate the fossil.

The chance of a linking form, which probably did not survive for any long period, but quickly passed into more stable types, meeting with such conditions is remote. Yet some such forms are known; and of many types a complete series reaching back to the beginnings of their history has been discovered; so that we are able to trace not merely the development of the type, but in some cases also its migrations. Occasionally a

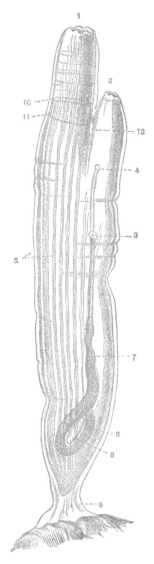

Fig. 7. *Ciona intestinalis*. The live animal seen in its test; some of the organs can be seen, as the test is semi-transparent. 1, mouth; 2, atrial orifice; 3, anus; 4, genital pore; 5, muscles; 6, stomach; 7, intestine; 8, reproductive organs; 9, stalk attached to a rock; 10, tentacular ring; 11, peripharyngeal ring; 12, brain.

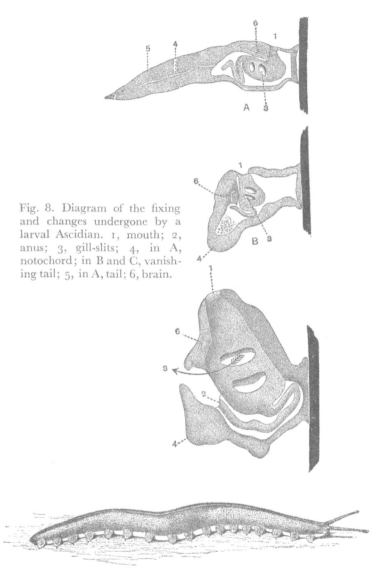

Fig. 8. Diagram of the fixing and changes undergone by a larval Ascidian. 1, mouth; 2, anus; 3, gill-slits; 4, in A, notochord; in B and C, vanishing tail; 5, in A, tail; 6, brain.

Fig. 9. *Peripatus capensis* × very slightly.

cataclysm will lead to the death and fossilisation of a large number of individuals together, as it did in the case of *Iguanodon bernissartensis* in Belgium; but generally the survival of isolated specimens is a matter of luck, like the stupendous luck which preserved two individuals of the ancestral bird *Archaeopteryx* in the lithographic slate at Solenhofen in Bavaria.

Thus, from the record of the rocks we can only expect three things so far as detail is concerned: first, that there shall be no case that tells against the theory of evolution —negative evidence that is of great positive weight when taken in connection with the rest; second, that fairly frequently we shall get series which link completely the living types with their primitive ancestors; and third, that now and again we shall have the luck to find a fossil which serves as a link between the larger groups. Besides these points of detail, we shall require to find that, taken as a whole, the series of fossils in the rocks should show a progressive increase in complexity of organisation, at any rate until one dominant type is re-placed by another; and that different types of organisa-tion should successively emerge. These must be our criteria, and all of them are satisfied.

Molluscs and crustaceans first supply the dominant races, then fishes, then reptiles, then mammals.

Within these groups, the earliest fossils are more generalised in structure than the later ones.

Complete series of forms are fairly plentiful. We may select the horse and the elephant as examples, but it must be understood that we might have taken many other equally good illustrations. The Ungulata had a slow-moving, marsh-loving, plantigrade ancestor called *Phenacodus*, with five-toed limbs (fig. 10). As time

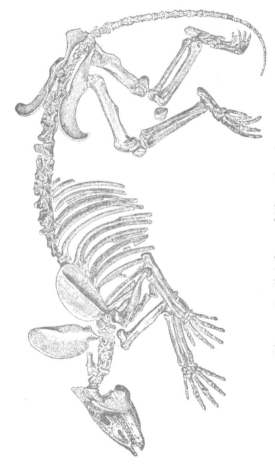

Fig. 10. Fossil skeleton of *Phenacodus primaevus.*

went on, they tended to acquire a better speed on drier ground by running on their toes. One group ran on two toes, the third and fourth: of these some, like the camel, lost almost all trace of the other digits; others remain still in a less specialised condition. The pig has the second and fifth toes nearly touching the ground; the hippopotamus has four well-developed toes, with a hint of the first as well (fig. 11). The other group took to running on the middle toe alone. Here too we find all stages among living types. The elephant has five toes, with the middle one a little enlarged; the rhinoceros has only three and a vestige; the horse has only one, with splint-bones representing two more (figs. 12 and 13). But the fossil horses show a very complete series from the Eocene onwards, starting with a fox-terrier sized horse with four toes on the hind-foot, through a larger one with three toes and a splint, a Pliocene one with one large and two reduced digits, to the modern horse.

So too with the elephants. In the upper Eocene we find *Moeritherium,* a tapir-like creature with a plastic nose and long incisors. In the lower Oligocene appears *Palaeomastodon,* with a long chin and the beginnings of tusks. He is followed by the long-chinned *Tetrabelodon,* a peculiarly ugly animal. The later Belodons found the long chin a nuisance; they shortened it and laid stress on the upper tusk. And so we arrive at the modern elephant, weak-chinned, but retaining the ancestral trunk and upper tusks, and even improving upon the ancestral skull (fig. 14).

So, too, the earliest fishes show a symmetrical, verte-brated tail (diphycercal), some later types develop an asymmetrical tail, with vertebrae running up into the

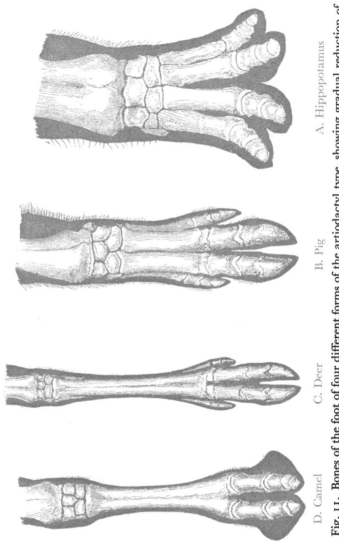

D. Camel   C. Deer   B. Pig   A. Hippopotamus

Fig. 11. Bones of the foot of four different forms of the artiodactyl type, showing gradual reduction of the number of digits, coupled with a greater consolidation of the bones above the digits. The series reads from right to left.

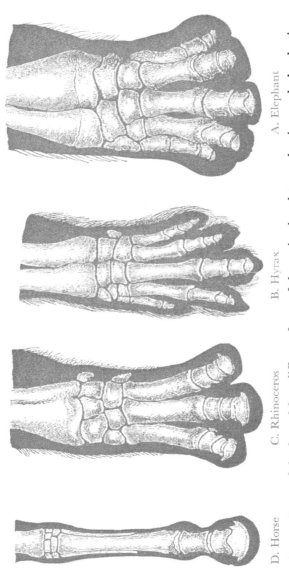

D. Horse   C. Rhinoceros   B. Hyrax   A. Elephant

Fig. 12. Bones of the foot of four different forms of the perissodactyl type, showing gradual reduction in the number of digits, coupled with a greater consolidation of the bones above the digits. The series reads from right to left.

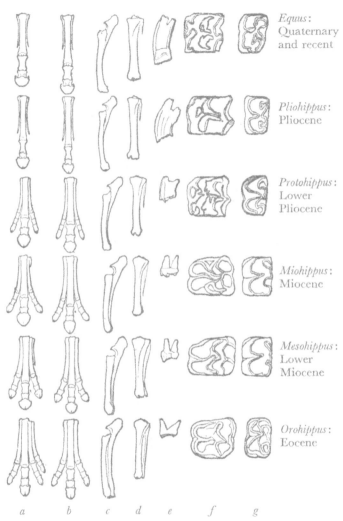

Fig. 13. Feet and teeth in fossil pedigree of the horse. *a*, bones of the fore-foot; *b*, bones of the hind-foot; *e*, radius and ulna; *d*, tibia and fibula; *e*, roots of a tooth; *f* and *g*, crowns of upper and lower molar teeth.

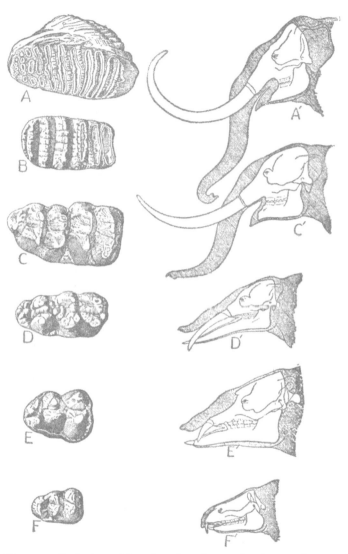

Fig. 14. Evolution of head and molar teeth of mastodons and elephants. A, A', *Elephas*, Pleistocene; B, *Stegodon*, Pliocene; C, C', *Mastodon*, Pleistocene; D, D', *Trilophodon*, Miocene; E, E', *Palaeomastodon*, Oligocene; F, F', *Moeritherium*, Eocene.

[27]

top lobe (heterocercal); while modern fishes are tending
to condense the end of their backbone into a stump from
which dermal structures radiate to support a symmetrical,
bifid, tail (homocercal) (fig. 15). A few fishes with the
primitive type of tail survive to-day; heterocercal tails
characterise the dogfish and his congeners; but homocercal tails, with their secondary symmetry, predominate.

Another example of the same kind may be used to
illustrate the existence of link-fossils. *Archaeopteryx*, the
ancestral fruit-eating bird, had teeth, three digits with
claws projecting from the wing, a number of reptilian
features in the skeleton, and above all a long vertebrated tail, consisting of some twenty vertebrae. Each of
these bore a pair of tail-feathers, of which the impres-
sions remain in the Berlin specimen (fig. 17). *Hesperornis*
and *Icthyornis*, later fish-eaters, still have teeth and traces
of a vertebrated tail (fig. 16); but the modern bird, like
the modern fish, finds that a stump from which the braking
and steering tail-feathers radiate, is far more efficient.
He has fewer reptilian features; though the legs are still
covered with unchanged reptilian scales, and it is
possible in the development of some feathers to detect
the process by which a scale turned into a feather.

Whether the birds originated from the reptiles along
the line of a swift-running dinosaur which took to
helping itself along by flapping its arms, or along the line
of an arboreal type which parachuted like a flying fox,
by means of a feathery structure involving the hind limbs
as well, is still uncertain. Embryological evidence faintly
suggests the latter, since quill-feathers are temporarily
formed on the hind-limbs as well as on the wings; but
another explanation of this is possible. We need not
discuss the evidence. It is clear that the reptiles had

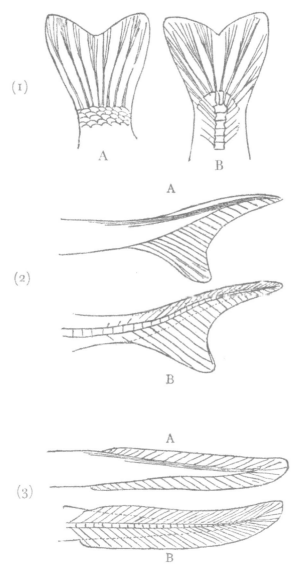

Fig. 15. (1) Homocercal tail; (2) heterocercal tail; (3) vertebrated but symmetrical fin (diphycercal). In each case (A) shows external and (B) internal structure.

already made a fairly successful attempt at aerial life,
for the pterodactyls were numerous. Arthropods, fishes,

Fig. 16. *Icthyornis victor.*

reptiles and mammals have all attempted to escape from
an overcrowded world by taking to the air, with some

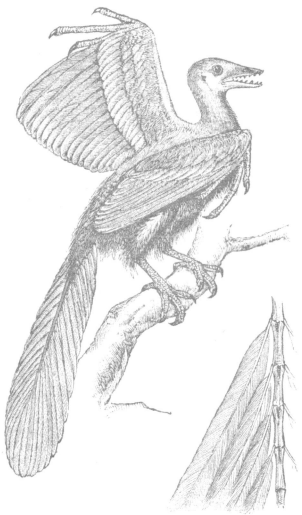

Fig. 17. *Archaeopteryx macura* and section of the tail.

degree of success, but only the insects and birds and bats have really made a good job of it.

The evidence of the fossils, then, is wholly in favour of evolution, and is incapable of any other reasonable interpretation.

Embryology. This may also be said of the embryological evidence. The law of von Baer, briefly summarised, states that each individual climbs its own family tree, as it develops. When a simple chalk sponge hatches out of the fertilised ovum, it segments, becomes a ciliated blastula,[1] invaginates into a ciliated, two-layered gastrula, and ends up in its final form, not fundamentally different from a *Hydra* without tentacles. The simplest Metazoa we know are more or less like permanent gastrulae. But the ancestral vertebrate also begins life in just the same way, if we may judge by *Amphioxus* (fig. 18). Blastula and gastrula are formed before the modifications involved in the possession of a dorsal nervous system, and an important intervening layer of cells, the mesoblast, which will give rise to coelom,[2] reproductive cells, muscles, and other things, complicate the development. Moreover, these same stages are perfectly recognisable in the higher vertebrates, though they are much

[1] A fertilised egg, if it has little yolk, divides rhythmically until it forms a hollow sphere, the *blastula*. One side of this is next squashed in, much as an india-rubber ball with a hole in it may be squashed in, so that a cup consisting of two layers of cells is formed. This is the *gastrula*. It subsequently elongates and becomes almost cylindrical. A small opening, the blastopore, is left. The whole structure is very like the cylindrical pond organism *Hydra*, without its tentacles.

[2] Essentially, a body cavity whose walls are intimately associated with the excretory and reproductive organs. There is another type of body-cavity in the insects and some other invertebrates which is part of the blood system.

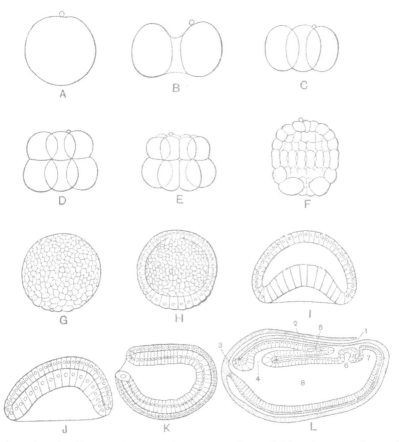

Fig. 18. *Amphioxus.* Segmentation; formation of blastula, gastrula and embryo. A–H, formation of blastula; I–K, gastrulation; L, diagrammatic longitudinal section of embryo. 1, neuropore—anterior opening of the neural canal; 2, neural canal; 3, neurenteric canal; 4, coelomic groove; 5, mesoblastic somite divided off from coelomic groove; 6, collar cavity; 7, head cavity; 8, alimentary canal.

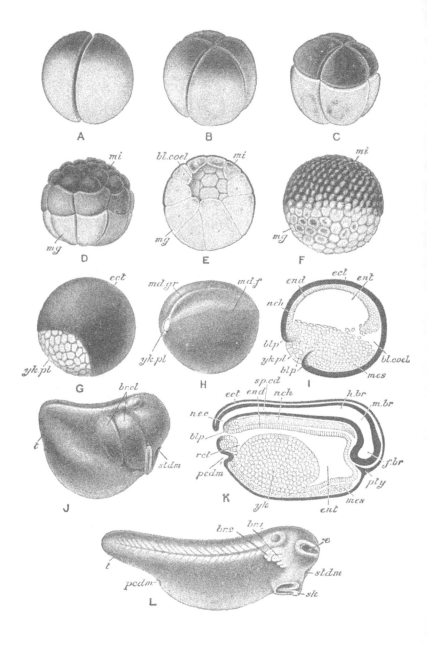

modified owing to the presence of yolk in the egg (fig. 19). In the Eutherian mammals (those which form a uterine placenta), even though the yolk is absent, development takes place exactly as if it were present, indicating the descent of these from a lower type which laid yolked eggs, as the duck-mole (*Ornithorhynchus*) does to this day.

Again, we have spoken of *Amphioxus* as probably not unlike the ancestor of the vertebrates, and have noticed the notochord and pharyngeal gill-slits. But every vertebrate embryo goes through a stage in which a noto-chord is the only skeleton; and moreover every verte-brate embryo has a tail and gill-slits like a fish, thus clearly indicating his descent from an aquatic ancestor. Early embryos of man, pig, chicken and fish are almost identical; only later do marked differences begin to arise (figs. 20 *a* and *b*).

Similarly, the feather-star is descended from the ancient and notable line of the sea-lilies, of which a few survive to this day. In a young stage the larva of the feather-star is nothing more or less than a tiny sea-lily (figs. 21, 22).

Fig. 19. Development of the FROG. A–F, segmentation; G, over-growth of ectoderm; H, I, establishment of germinal layers; J, K, assumption of tadpole-form and establishment of nervous system notochord and enteric canal; L, newly hatched tadpole. *bl.coel.* blastocoele; *blp.*, *blp'.* blastopore; *br.* 1, *br.* 2, gills; *br.cl.* branchial arches; *e.* eye; *ect.* ectoderm; *end.* endoderm; *ent.* enteron; *f.br.* fore-brain; *h.br.* hind-brain; *m.br.* mid-brain; *md.f.* medullary fold; *md.gr.* medullary groove; *mes.* mesoderm; *mg.* megameres; *mi.* micromeres; *nch.* notochord; *n.e.c.* neurenteric canal; *pcdm.* proctodaeum; *pty.* pituitary invagination; *rct.* com-mencement of rectum; *sk.* sucker; *sp.cd.* spinal cord; *st.dm.* stomadaeum; *t.* tail; *yk.* yolk-cells; *yk.pl.* yolk-plug. D and E show clearly the blastula-stage, H and I the gastrula.

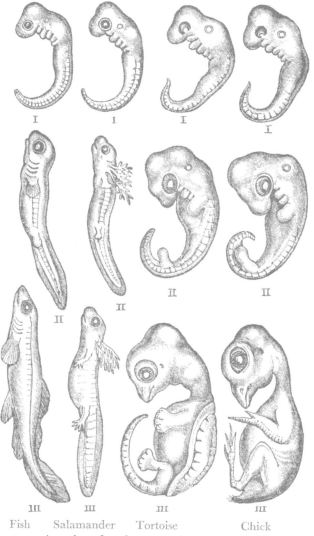

Fish   Salamander   Tortoise   Chick

Fig. 20 *a*. A series of embryos at three comparable and progressive stages of development (marked I, II and III), representing each of the classes of vertebrated embryos below the Mammalia.

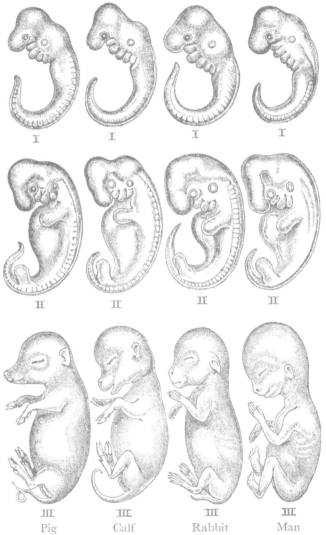

I   I   I   I

II   II   II   II

III   III   III   III

Pig   Calf   Rabbit   Man

Fig. 20 *b*. Another series of embryos, also at three comparable and progressive stages of development, representing four different divisions of the class Mammalia.

We have already noticed that the sea-squirts begin life in a shape suggestive of a small tadpole: indeed the resemblance is close. It is only later that they stand on their heads and assume their uncouth maturity.

The worms and the molluscs start as *Trochophore* larvae, very like the coelenterates known as Ctenophora, except that their one mouth-opening becomes divided into two—a change still recognisable in the frog embryo (fig. 19, K), where the blastopore becomes narrowed in the middle, leaving a tiny communication with the archenteron at the top, and a dimple that becomes the anus below. (The true mouth is a new structure.) Then the worm elongates and segments the lower half of the body, while the mollusc grows a shell on the middle. But the members of both phyla start life as *Trochophore* larvae (fig. 23).

We have already spoken of the Crustacea, where forms so different as the branching tubes of the *Rhizocephala* and the shell-covered barnacles, both begin as typical *Nauplius* larvae by no means unlike tiny trilobites of the Devonian in general appearance (figs. 24 and 25). Instances might be multiplied indefi-

Fig. 21. *Rhizocrinus.*

nitely; but those we have already given will suffice to show that no other interpretation but an evolutionary one can give a reasonable explanation of the phenomena of embryology: a conclusion which a study of detail only renders the more certain.

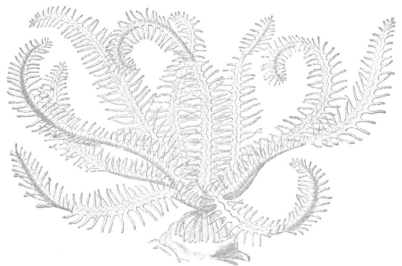

Fig. 22. Antedon. Side view of entire animal.

Geographical distribution. The first fact that strikes the investigator is that of surviving types, those which are the earliest and most generalised have the widest distribution. *Peripatus* occurs in South America, South Africa, Malaya and Australasia. The lung-fishes are found one in Brazil, one in Africa, and one in Australia. Yet the commonest of more recent types are comparatively local. There were no European rats and mice in America: America has her own species, which differ from ours in a few details. There were no rabbits in Australia

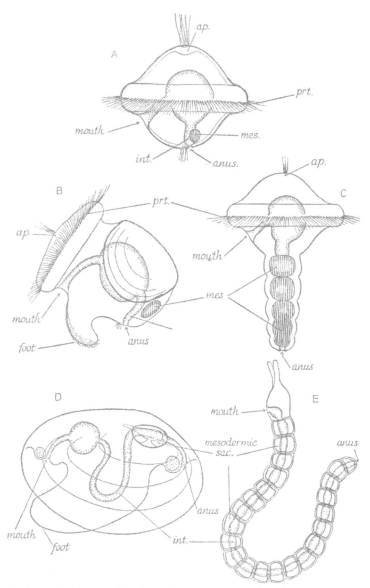

Fig. 23. Diagrams showing the structure of the *Trochophore* and its passage into the annelid and mollusc respectively. A, the *Trochophore*; B, a molluscan larva; C, a worm-larva; D, a young bivalve mollusc; E, a young worm. *ap.* apical plate; *int.* intestine; *mes.* mesoderm; *prt.* prototroch.

until we were foolish enough to put them there; thereby demonstrating that it was not at all a question of an

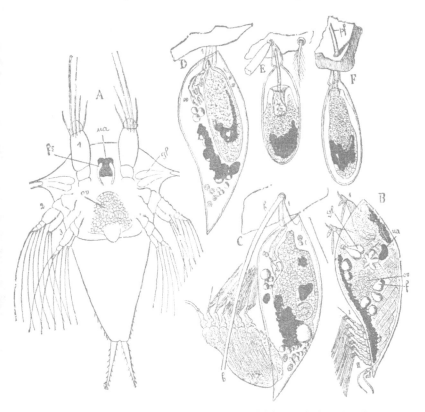

Fig. 24. *Nauplius* and *Cypris* larva of a Rhizocephalan showing attachment and degeneration. Compare fig. 3 (*Sacculina*).

unsuitable climate. There were no sheep nor cattle either; and indeed no mammals at all, except the primitive marsupials, and the still more primitive mono-

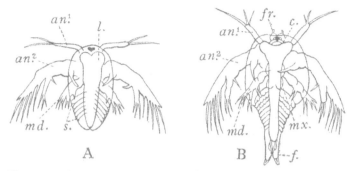

Fig. 25 *a*. Two stages in the development of *Apus cancriformis*. A, *Nauplius* stage at the time of hatching; B, stage after first ecdysis. *an.*¹ and *an.*² first and second antennae; *md.* mandible; *mx.* maxilla; *l.* labrum; *fr.* frontal sense organ; *f.* caudal fork; *s.* segments. Compare this typical Nauplius with the trilobite below.

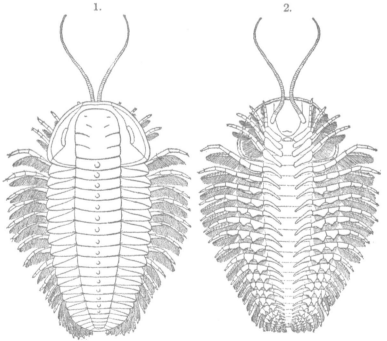

Fig. 25 *b*. Trilobite, *Triarthrus becki*, restored. 1, dorsal; 2, ventral aspect. Twice natural size.

tremes (and the dingo, if it was not introduced by man);
for the simple reason that the land bridge between
Australia and the mainland, where the later types were
developing, had sunk before they got far enough to
arrive. There are no humming-birds in the eastern
hemisphere, but their place is taken by the sun-birds of
Japan.

On the mountains we find little islets of arctic fauna
and flora, survivals of the time when arctic conditions
were widespread.

Shallow-water fauna are distributed roughly in parallel
belts running north and south with the contour of the
continents. The shallow-water fauna on the opposite
sides of the Isthmus of Panama, for instance, are very
different. Similarly the land fauna on the two sides of
a mountain range are frequently quite distinct. The
snails in the valleys of Hawaii which run down to the
coast, with arid regions between, show progressive
differences; so that a snail from a distant valley is far less
like the one under consideration than a snail from the
next valley.

Such facts—and you find them wherever you look—
suggest that barriers to dispersal result somehow in the
production of a really distinct fauna.

What constitutes a barrier to dispersal depends on the
nature of the animal. Dry land is usually a barrier to
fishes, though an eel will cross quite large spaces. Deep
sea is a barrier to shallow-water forms. To land animals a
wide space of sea is an effective barrier, though birds may
traverse it under certain conditions, and insects may be
blown across, or drift on logs, and seeds may float in the
water and arrive uninjured. (As Darwin showed, vast
numbers of seeds are carried in the mud on the feet of

wading birds.) A mountain range may be a barrier even to birds, or to strong mammals, not because it is in itself impossible to cross, but because the cold does not allow the plants or smaller creatures which serve as their food to exist. If an animal finds its food getting scarcer, it will go no farther.

Geographical distribution definitely suggests the idea that animals radiated from various centres of dispersal, and that barriers of various kinds determined the actual routes and limits of that dispersal.

A study of the fauna and flora of oceanic islands illustrates the essential points with remarkable clearness. We may follow Romanes (*Darwin and after Darwin*) in a brief study of a few such islands. Even though his actual figures may be out of date, later investigation cannot materially alter the broad aspect of the facts.

Oceanic islands are, for our purposes, islands, far removed from continental land, which have existed since a fairly remote geological epoch. We may first consider the Galapagos Archipelago, the Sandwich Islands, and St Helena.

The Galapagos form a group about 500 miles west of America, almost on the equator, in the region of perpetual calms. In these islands the giant-tortoise still lives, a different species in each island. They cannot pass across the water, and isolation leads to their developing in different directions. Giant lizards also survive, some 4 feet long, but as these have taken to browsing on seaweeds they are quite happy in the water, and consequently we find the same species throughout the islands. For the rest, all the molluscs, insects, and all the birds except one are peculiar to the islands, as are the ten reptiles. The one exception is the universally distributed

American rice-bird. No molluscs, reptiles, birds or insects that were not peculiar, with this one exception, were known to Romanes.

In the Sandwich Islands we find the same story. All the numerous molluscs, both the reptiles, and all the birds are found nowhere else.

St Helena, 1100 miles from the nearest land, was once virgin forest, which must have been full of fascinating survivals from an earlier epoch; but the island was stripped nearly bare in the course of two hundred years, and little of the fauna survives. Of the remnants all the insects, all the molluscs, and the one bird are peculiar. There are no reptiles; and, as in the other islands, no mammals. Thus in these islands out of 658 animals, only one bird is found anywhere else in the world.

Two-thirds of the plants also are peculiar: a most striking thing when one remembers the much greater chance of transport of a plant.

In the British Isles, recently separated from the continent by a narrow strip of water, we find only about 1 per cent. of the flora and fauna peculiar; and such peculiar species as exist are closely allied to the continental forms; whereas in the oceanic islands many of them belong even to peculiar genera.

Thus it becomes clear that an explanation which bases this distribution on the idea of dispersal which is checked by some barrier, so that the isolated colony pursues its own line of development, and sometimes preserves ancient types, is the only one which can satisfy our reason.

Some oceanic islands which at first seem to controvert this idea are found, on further investigation, to support it. Thus, the Azores have a far less peculiar fauna and flora than the Galapagos, though there are many peculiar species. Yet they are farther from continental land.

These islands are, however, in the region of trade winds, and storms, so that the chances of accidental transport are greatly increased, and we should expect to find many European and African types. Birds and flying insects may easily be blown across; and the chance of driftwood carrying small creatures to the shore of the islands is not remote.

Thus we see that the argument from distribution in space, like that from distribution in time, is wholly in favour of the evolution of types radiating from various centres and being blocked or diverted by barriers to dispersal. Any other explanation is irrational.

Variation under domestication. When we further take into account the undeniable fact that even in the short time, geologically speaking, which has elapsed since neolithic man began the arts of agriculture and breeding, the human race has produced many new types under domestication, the argument in favour of evolution grows overwhelming. In some cases we know the original. All the strains of wheat were probably derived from the inadequate plant which still grows wild on the slopes of Mount Hermon. We know the ancestral Blue Rock from which all the Pouter, Fantail, Jacobin, and Tumbler Pigeons are derived. We know the Malayan Jungle Cock from which we get Silkies and Houdins and Wyandottes, and the rest.

The wild boar is with us still, as well as his obese derivative, the prize pig.

In other cases, though we are perfectly certain that they are the work of man, we do not know for certain whence our modern breeds spring. We see breeders still producing new types of dogs; but whether our Bulldogs, Pekes, Newfoundlands and the rest (types so different that if they occurred wild we should be inclined to give them specific, or in some cases generic rank) sprang originally from some kind of dingo, or other wild dog, or from the wolf or jackal, or from some hybrid we do not know. Very likely they originated in the domestication of more than one kind of beast. What we know certainly is this; that at times freak-individuals are born, which differ markedly from the type. Some of these man has perpetuated by a process of selection. Such freaks, which occur fairly freely under natural conditions, are known as mutations. They may or may not have played an important part in evolution; in the discussion of evolution at least they have done so. We shall return to them shortly.

Chapter II

THEORIES OF EVOLUTION

Having outlined briefly some of the lines of evidence which have driven all those who are competent to judge and weigh this evidence to an inescapable acceptance of the fact of evolution, our next business must be to summarise the chief attempts that have been made to construct a theory which will be consonant with, and explain, these facts.

Such theories really fall into one of two classes, the first associated with the name of Lamarck, the second with that of Darwin.

It is probably not an exaggeration to say that there is literally no question of more importance to the physical well-being of the human race than the decision between the rival claims of these two views. They underlie all attempts at ameliorating the lot of men, whether we are aware of the fact or no. At present man's politics are largely a matter of prejudice; but the effects of his legislation are rigorously determined by biological laws. The sooner these laws are understood, and the sooner reason takes the room of prejudice, the better for the human race. Unfortunately in some of the most important matters the right answer is still in doubt. Some biologists have gone beyond their brief. Nevertheless it will be our business later to point out that hesitation where there is no absolute certainty is perfectly compatible with strong action where certainty does exist.

We will begin by stating in this chapter the actual issues between the theories of the Lamarckian and Darwinian types. Our later studies will add much which affects the details of the problem and sheds new light; but the main issues still remain clear-cut. Though Darwin would have been the first to modify some of his suggestions had he known what we know, his name may still fitly stand for one type of explanation; even though we may call it "neo-Darwinism", and feel that some forms of it are more "new" than Darwinian.

The singularly accurate guesses of some of the Greek philosophers, from Anaximander onwards, may be dismissed as belonging rather to the realm of speculation than that of science.

Lamarck in 1801–9 published a theory of evolution based on the idea of use-inheritance. He pointed out that an organism, by using some particular organ, developed it appreciably. This is obvious fact. But he further added that this increased development would be transmitted to the offspring. Naturally it would not develop without the stimulus of environment: nothing does so, though it is not only the extreme Lamarckian who holds that in time its development may be determined solely by internal factors, and thus appear to be automatic (for after all, a vertebrate embryo does develop a tail while still in the egg); but the central Lamarckian idea is that, given the same environmental stimulus, the offspring will even improve upon the parent.

Such a view depends upon what is loosely called the inheritance of acquired characters, an acquired character being, in the sesquipedalian but exact terminology of the zoologist, an exogenous somatic modification —that is, a modification of bodily structure produced by external influences.

A few illustrations will make the meaning clear; but the vital question is, of course, whether we have any evidence that such characters are inherited.

Suppose an antelope took to stretching up and browsing on the lower branches of trees. His neck would grow a little longer. His children would be born with longer necks, and would still browse on trees; and their children's necks would be longer still, till ultimate great-great-grandchildren would be giraffes. This is perhaps an extreme example.

The wading birds, again, may be supposed to arise through a tendency in a bird experimenting on a water-edge diet to stretch up so as to avoid getting his feathers wet. The captious might contend that he was likely to give up the effort before he had stretched far! Are the children of blacksmiths born with horny hands? Most certainly they are not, though the trade has often passed from father to son through many generations.

But there are much stronger cases than these. The flat-fish start life as symmetrical little creatures. One side puts on more flesh than the other, and the fish begins to fall over sideways. An eye scraping along the sand would be of no use, but rather a disadvantage. By a modification of the muscles, accompanied it is said by actual muscular effort, the eye is twisted over, and in due course is established the adult deformity of both eyes on the same side. This case certainly looks like an instance of use-inheritance.

On the other hand it is quite certain that the complicated spiny shells of many molluscs are derived from simple structures of the winkle type. Here is a case of evolution where by no conceivable stretch of imagination can the idea of use-inheritance be taken as the explanation. No direct activity of the mollusc could alter the spines on its shell, which is simply a house.

We shall have much more to say of Lamarck's theory.

Whether it is true or false, we may note that it was first put forward without any direct evidence save its superficial reasonableness; and in spite of a considerable amount of experimental work the last 130 years have produced no evidence which is held to be convincing by the majority of zoologists.

The outstanding objection has always lain in the difficulty of conceiving a mechanism by which a modification of some outlying region of the body should so affect the germ-cells that exactly the same modification should in due course appear in the offspring. These might easily be rendered stronger or weaker; but imagination boggles at anything more specific than that. It may be said in anticipation, however, that recent work does in some degree lessen this difficulty.

The great merit of Darwin's work was, that, before its publication, nineteen years were spent in the collection and examination of a vast amount of evidence; that it moved in logical sequence; and that each stage of the argument was derived from the study of concrete facts. In this it differed from the theory of Lamarck, which amounted to little more than a brilliant guess.

Darwin started from the fact of the nexus of living things. Every living thing is a factor in the lives of others, constituting an important part of some other organisms' environment.

All the creatures which constitute this warp and woof of life produce far more young than can survive. Thus, the elephant, which is the slowest breeding of animals, would at the end of 750 years have 19,000,000 living descendants. The thrush is not a specially prolific bird. It has two broods a year, and may breed for nine years, laying about four eggs in each clutch. If all the offspring lived and mated, at the end of the first thrush's life he would have about 19,500,000 descendants. Twenty years later there would be 1,200,000,000,000,000,000,000. If all these thrushes stood shoulder to shoulder about one in every 150,000 would find perching-space on the whole surface of the terrestrial globe (MacBride).

When one turns to organisms which really reproduce rapidly, the figures grow rather staggering. Rotifers are only just visible to the naked eye, yet Professor Punnett calculated that if he had been able to rear the whole of the rotifers he was breeding, theoretically he would obtain from the original parents, from sixty-seven generations in less than a year "a solid sphere of organic material with a radius greater than the probable limits of the known universe". With *Paramoecium* breeding at the normal rate of three bi-partitions in forty-eight hours Woodruff estimated that, theoretically, in fifteen years the descendants of one individual would spread beyond the confines of the known universe, and the margin of the mass would be pushing out into space with the velocity of light. When we remember that some fishes lay 30,000,000 eggs a year; that a queen white ant may lay 80,000 eggs a day; and that flourishing bacteria may undergo fission every ten minutes or so, there is little need to stress Darwin's point that there must be a struggle for existence.

Such a struggle involves many factors, such as the obtaining of food, finding a mate, and coping with the general conditions of environment. From such considerations Darwin concluded that even a slightly better adaptation to environment, or a slightly increased attractiveness to the opposite sex would give the individual a better chance of surviving and propagating.

The nature of this struggle deserves a little consideration. It is not as a rule a fight to the death between different animals, though frequently a limited food-supply, for example, means that there is not room for more than a few species to survive, and the types best adapted to all the many conditions obtaining in that region will oust the others. Thus Darwin[1] pointed out that in Paraguay neither horses, cattle nor dogs run wild, though they are plentiful both to the north and south of the district. The cause is that a certain fly lays its eggs in the navel of the new-born offspring. He adds, "The increase of these flies, numerous as they are, must be habitually checked by some means, probably by other parasitic insects. Hence, if certain insectivorous birds were to decrease in Paraguay, the parasitic insects would probably increase; and this would lessen the number of navel-frequenting flies—then cattle and horses would become feral, and this would greatly alter (as indeed I have observed in parts of South America) the vegetation: this again would largely affect the insects; and this, as we have just seen in Staffordshire, the insectivorous birds, and so onwards in ever-increasing circles of complexity". Though better examples might be cited, this quotation does illustrate both the nexus of life and the struggle for existence between different species.

Darwin goes on to point out that between species of the same genus having generally similar habits and constitution, the struggle is still more severe. Thus the missel thrush ousts the song thrush in Scotland, the brown rat ousts the black, the eastern cockroach ousts the larger species, and so forth. And this leads him on

[1] *Origin of Species*, ch. III.

to his central doctrine that within the species itself the competition is most severe; and that in consequence not only do the "fittest" species survive, but the fittest members of a given species. Thus he comes to modification of species by natural selection, and his theory of evolution.

Before proceeding further it may be well to set down clearly the main stages of Darwin's argument, since they are often but vaguely grasped. Starting with the fact of the *over-production of individuals* ("Even slow-breeding man has doubled in twenty-five years, and at this rate, in less than a thousand years, there would literally not be standing-room for his progeny"), he points out that there must be a *struggle for existence* between those individuals as well as between different species. Since there is *variation,* and no two individuals are exactly alike, some are bound to be better suited than the rest to the conditions of their life; and these will be more likely to survive. The process which leads to such *survival of the fittest* Darwin called "natural selection", a term which at once recalls the analogous *artificial selection* practised by breeders who wish to modify the type of their stock. The result will be that on the whole it is these *fittest* who will survive to mate together; and since *heredity* shows that offspring tend to be more or less like their parents, the next generation will vary, not about the mean of the original stock, but about the mean of the surviving stock. In this way Darwin believed that the type of the race would shift gradually in the direction best suited to survival, and the species would be modified.

Among many examples of such modification of species Darwin quotes two races of wolves in the Catskill

Mountains, "one with a light, greyhound-like form, which pursues deer, and the other more bulky, with shorter legs, which more frequently attacks the shepherds' flocks"; and he illustrates the possibility of advantage arising from small modifications of an organ by the fact that the hive-bee can easily get the honey from the clover *Trifolium incarnatum* while it cannot do so from the closely similar *Trifolium pratense*, whose corolla-tube is of almost the same size, but whose harvest is reaped by the humble-bee. The third and fourth chapters of *The Origin of Species* are full of interesting material, which may be consulted with advantage.

The next factor to be considered, then, is that of heredity. Like father, like son, is proverbially true; yet no two individuals are indistinguishable. The most like are those known as "Philippine" or identical twins. These are the product of a single fertilised ovum of which the halves have become accidentally separated after the first cleavage. When the separation is incomplete we get monsters with two heads, or, in more complete cases, "Siamese Twins"; complete separation results in two whole individuals who are extraordinarily alike and of the same sex. This likeness extends to mental characteristics: however the twins are brought up, even if it be in different hemispheres, their abilities are the same, and tend in the same direction. Yet, like as the twins may be, they are not absolutely indistinguishable.

In normal cases the children, even ordinary twins, show decided differences from one another and from their parents. The fact of the existence of variation is as certain as the fact of heredity. Children tend to be like their parents; but they also tend to differ from them. Variation is a fact which cannot be overlooked.

It was upon the nexus of life, heredity, variation, and the struggle for existence that Darwin founded his theory, and we may again stress the fact that the popular conception of this struggle as "eat or be eaten" is to a large extent a false one. The struggle is mainly between individuals of the same species. If you plant one square yard with seeds all of the same kind, and another square yard with different seeds you will probably get many more plants developing in the second case, for the demands of different plants are different. The same holds good of animals. It is all a question of food-supply and the like.

Let us take an illustration. Suppose we consider an organism which floats in the surface-waters of the sea. No doubt whales and many other creatures are undiscriminating; but none the less, the more transparent the organism, the greater chance it will have of escaping the jaws of some other organism less wholesale in its methods than a whale. As time goes on, more of the transparent ones, and fewer of the opaque will be left; and thus the chance of two transparent ones meeting will increase and some of their offspring are likely to be more transparent than either parent. Eventually only transparent ones might survive.

The giraffe gives us a more precise illustration. Darwin would say that chance variation produced some antelopes with longer necks, and with an aspiring mind which led them to browse on trees. This new food supply gave them additional vigour and a new source of supply, especially valuable in times of dearth. Thus the long neck has a definite survival value. We may perhaps add to this preferential mating, long-necks tending to mate together; for Darwin laid stress on sexual selection.

On either hypothesis, Lamarckian or Darwinian, the result in this case is the giraffe; but the one explains it by the inheritance of *stretched* necks, the other by the survival-value of chance variations in the direction of long necks, and the fact that offspring tend to vary on either side of the parental mean to about the same extent, so that the average of the offspring of long-necked parents will tend to have longer necks than the average of those born from short-necked parents.

Two obvious difficulties which confronted the views of Darwin were: (1) that the type of variation (continuous variations) on which he relied exhibits very small differences—too small to have a survival value; a half-inch or less would not help our antelope much in nibbling trees; (2) that such a theory would only account for the evolution of characters which had a definite adaptive value. The tail of a peacock both hampers his flight, and is a drain upon his energy.

The discovery of discontinuous variations, largely due to Bateson's work, and (by de Vries) of the particular type known as mutations, removed the first difficulty to some extent (though we shall see that some doubt has been cast, not altogether convincingly, upon this).

The work of Johannsen on "pure lines" (discussed below) finally disposed of the idea that continuous variation played any part in evolution; and some statistical work of Karl Pearson confirmed this from another side.[1] Thus we are left with the dilemma that either discontinuous variations lie at the bottom of evolution, or else that we must accept the Lamarckian explanation; unless indeed we decide that both of these two have played their part.

[1] See p. 81.

Johannsen took a bag of mixed kidney beans, some large, some small, and sowed them. As might have been expected, on the average the plants grown from large beans gave, when the fruit ripened, larger beans than did the plants grown from the small beans. But in a given pod the beans always tend to be larger in the middle of the pod. Taking now large and small beans from the *same pod* he found that the small ones gave as good an average size as the large ones. Thus he showed that the small "continuous" variations on which Darwin relied had no evolutionary value. Johannsen has proved that selection within what he calls the "pure line" is of no use at all: only in a mixed population does it avail. In other words, the smallness of a bean may be due to one of two causes. Either it comes from a small-beaned race, or its smallness is due to the chance of its position in the pod. Only by breeding can you decide which of these is the operating cause.

The second difficulty was in part disposed of by Darwin himself in the theory of sexual selection.

The peacock, the lyre-bird, the bird of paradise, the Argus pheasant, to take only four examples, all show a marked sexual dimorphism. The male is far more gorgeous than the female. Moreover, at the mating season he postures in front of the desired mate, and doubtless the handsomest males get the best wives.

So too the deer use their antlers to obtain the best mate, both by display and by fighting. The mandrill woos by displaying alternately his blue face and his scarlet backside. The spider dances a kind of Russian ballet before his mistress, who in some cases ungratefully eats him when she is tired of him. The bower-bird builds a pleasaunce for his inamorata, planting flowers and bright pebbles, and erecting a kind of tent in the midst with two doors, through which the birds chase each other in and out.

Indubitably we have here evidence of the beginnings of the aesthetic sense, destined to such noble transformation, as well as the explanation of the survival of cumbrous appendages. After all, the peacock which can grow a specially good tail must be a fine specimen; so the offspring will start with a good inheritance. That something which begins as a method of attracting a mate should end as the basis of all art should not surprise us: we frequently meet with the complete transformation of the physical function of an organ; it is interesting, but not unexpected, that the mind-functions should show a similar phenomenon. The evolution of mind is as much a fact as the evolution of body, and, as every student of psychology knows, von Baer's law holds good in this region also.

Isolation, whether physiological (due to mating-preference or group sterility) or geographical, is another important factor, which we have already noticed in the Hawaiian snails. If we imagine a herd divided by some chance into two, it is pretty certain that after a few dozen generations there will be a marked difference between the two herds. The two halves are unlikely to carry identical hereditary factors, and selection will work along slightly different lines. Weismann laid down the canon that what he called panmixia—indiscriminate breeding—alone will keep the type uniform; and though no doubt he unduly stressed the idea, it is broadly a true one.

Another very important point that has gradually emerged is that a change in one organ is frequently, perhaps usually, correlated with changes in other organs.

For instance, the adoption of an upright gait does not involve the legs only, but very many organs of the body. Similarly, a very small change in the metabolism of, say, a butterfly, may completely alter its coloration, and even its venation. Thus a useful change may bring in its train other changes, some of which may not be useful at all. If they are definitely harmful, they will of course be eliminated by natural selection.

Nevertheless Darwinism has always been haunted by the spectre of utility. Adaptive evolution must always be based on the selection of the useful; yet on the one hand it is hard to be certain that all structures which survive are useful, or else correlated changes; and on the other hand it is quite certain that the adaptations which the individual makes to his own conditions are extremely useful, though according to the Darwinian theory they are not inherited.

As regards the first point, there is certainly much to be said for Lloyd Morgan's doctrine of emergent evolution—the phenomenon of the emergence of new qualities in the evolution of the universe. Though not an explanation, it seems to be a fact; a fact, however, which itself requires explanation; and such an explanation will inevitably take the mind into the region of philosophy. But equally it is a fact which must not be neglected when we are thinking of the immediate utility of some characteristic which arises out of the blue. Yet, as a theory, emergence is unsatisfactory. According to it something emerges out of nothing; and something from nothing is as hopeless a doctrine philosophically as something for nothing is economically. New qualities do emerge; what we have to ask is *why* they emerge; and Lloyd Morgan's view affords no real answer.

Mimicry again offers an instance of the difficulty of the utilitarian point of view. Though there are many cases of a mimicry which is undoubtedly protective, either through making the organism invisible, or in making it resemble some harmful or unpleasant type, this explanation will not cover all cases. Often the mimic is more plentiful than the model, so that selection cannot be supposed to work; in some instances the model is either rare or non-existent in that region, as in the case of some of the hover-flies of Japan which mimic bees and wasps. Punnett has worked out the case of *Papilio polytes* and other butterflies which are said to be mimics, and the difficulties which he presents to the simple theory are formidable.

As regards the second point, it is certainly a regrettable fact that children are not born knowing the Latin grammar or the differential calculus, any more than blacksmiths' children are born with horny hands, or the children of wine-tasters with an educated palate. But the real question is whether such children acquire these characteristics more readily than the ordinary run of children; and if they do, whether this is due to the past activities of their parents or, as is far more probable, to a hereditary tendency operating in parents and children alike.

But it must be remembered that Darwin did not slam the door on the Lamarckian factor, as the neo-Darwinians have done, leaving all to blind chance and natural selection. His "provisional theory of pangenesis" suggested that possibly there might be small corpuscles which he called "pangens", scattered throughout the body, which might be modified by what happened to the particular organ in which they were, and which might migrate to the germ-cells, and thus influence the offspring. It was a mere guess, untenable in that form, but it did foreshadow an idea which has since received some experimental justification, as we shall discover later. It is now just conceivable, though completely unproven, that education, manual or intellectual, might gradually impose upon the germ-plasm of the race a greater educability; but the balance of evidence is overwhelmingly against such an idea.

In our study of heredity we shall see that many new facts, unknown to Darwin and Lamarck, have to be taken into account. But it is not necessary to discuss here theories, such as that of orthogenesis, which have been advanced to account for phenomena like the fatal over-development of some feature which leads to extinction; we shall fulfil our object better by concentrating on the main issue. What matters to us is whether Lamarck or Darwin was on the right track. For the moment we will confine ourselves to some recent arguments about the inheritance of acquired characters.

Chapter III

THE INHERITANCE OF
ACQUIRED CHARACTERS

Towards the end of the nineteenth century Weismann, compelled by eye-strain, I believe, turned his attention from cell-work to the theory of evolution. His contribution eventually became an elaborate theory, of which only the starting-point need concern us. Weismann pointed out that in every generation the body was formed afresh from the zygote, itself the result of the union of two germ-cells. These germ-cells were recognisable at least at a very early stage of development as different from the other cells, and they alone could be concerned with heredity: the body-cells passed nothing on but simply died, while the germ-cells were potentially immortal (fig. 26). Each organism alive to-day is alive in virtue of the germ-substance which has been handed down through countless generations. He therefore drew a firm line between the germ-plasm and the somatoplasm, as he called them. The germ-plasm is immortal; the somatoplasm is born again from it in each succeeding generation.

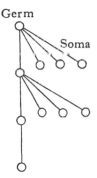

Fig. 26.

In this doctrine of the continuity of the germ-plasm, and its absolute isolation from somatic influences, Weismann no doubt went too far; but he at least propounded a logical view which has not yet been satisfactorily countered. Nevertheless we must not forget that the germ-plasm is being continually reinforced; and that it is not logically inevitable that germ-plasm should be incapable of building up the material it assimilates into something slightly different. Continuity does not in fact necessitate unchangeableness.

Weismann's hypothesis was destined quickly to find support from two sides. On the one hand, the study of nuclear division seemed to offer incontrovertible evidence that hereditary characters were carried by the chromosomes, and that, broadly speaking, these preserved each its own identity through all the changes and chances of nuclear division and the merging process of the resting-stage. On the other hand, the rediscovery of the work of Mendel on heredity, and its development during the first decade of the twentieth century, gave an independent buttress of great strength. It seems as though the somatoplasm must perish and leave no memorial; in which case, good-bye to the theory of Lamarck.

Let me remind you once again that this is no academic question, but one fraught with untold consequence for the human race. The whole structure of modern democratic ideas is founded unconsciously upon the Lamarckian view.

Even in the strictest sect of the zoological Pharisees, in which I was myself trained, it was of course admitted that unhealthy conditions might weaken the germ-plasm, and result in feeble offspring; and that good conditions might produce a stronger germ-plasm; but the idea that by altering the environment in any way you could change the actual hereditary characters was strenuously denied. The exogenous somatic modification (acquired character) could not repeat itself in intensified form in the offspring, or indeed in any form at all; even if the same environmental stimulus was supplied, that stimulus would produce on the average only the same effect in the offspring that it produced in the parents. Without the stimulus, of course, the character could not develop at all.

Since those days evidence has accumulated which tends to make the mind less certain. This evidence is as yet far from conclusive, in the eyes of most zoologists; and the evidence of the botanists has hardly received the careful scrutiny which it deserves; but there is enough to make us hesitate to decide in favour of the absolute isolation of the germ-plasm. At present I should decline to go farther than this in the way of positive affirmation; and I still find myself entirely unable to adopt anything resembling the crude Lamarckian hypothesis.

It is fair to say, however, that there has always been a tendency for zoologists to overlook the fact that, while admittedly the chromosomes of the nucleus carry the hereditary factors, both the egg and the sperm consist of *more* than chromatic nuclear material. Each germ contains some cytoplasm in the cell, more or less modified: the egg contains a considerable amount. Further, eggs and spermatozoa are bathed in the secretions of Cowper's glands, and the prostate gland, in mammals, and this may mean the intrusion of material neither nuclear in origin nor even belonging to the germ at all. Besides this, the ovary and testis receive a blood supply which must carry many hormones, and these may have their effect.

Some five and twenty years ago it was shown by Towers that if female chrysomelid beetles were subjected to abnormal conditions while the eggs were maturing, mutations were produced. MacDougal showed that the injection of chemicals into the ovaries of the evening primrose also produced mutations. These mutations were inherited, but showed no adaptive features. More recent work is summarised in MacBride's article in *Evolution in the light of Modern Knowledge*, from which we may extract some information, and in J. Arthur Thomson's *Concerning Evolution*.

Tornier studied the freak goldfish so dear to the Chinese. The common goldfish, before the gold or silver colour is established, is a very ordinary fish. The Chinese keep these fish in small pots full of foul water, and in the half-dark; and the results are the Meteor Fish, in fair round belly, whatever it may be lined with, and the Celestial Fish, two-tailed, with drooping fins and telescopic eyes.

Under the conditions of lack of oxygen and foul water the development of the upper part of the egg, which forms the fish, is delayed, while the yolk swells as usual under the influence of the water. The pressure of the yolk inside the tough egg-membrane disturbs the development of the embryo resulting in these strange heritable mutations.

The obvious conclusion is that these mutations are evidence of unhealthy stock; so feeble that they could only survive under special care. De Vries noticed that many of the mutants of his evening primroses were very unhealthy also; and the same is true of the mutations of *Drosophila*, the banana fly, in captivity. The conclusion drawn is that mutations cannot be the basis of evolution, as the neo-Darwinians hold, since they could never survive in competition with normal forms. We have already had to abandon Darwin's own idea that continuous variations lie at the bottom of evolution: it is suggested that we must also abandon discontinuous variations.

This argument seems to me to go too far on insufficient premises. What the Lamarckians ought to establish is that vigorous mutations are *never* formed; and this they have made no attempt to do. On the face of it one would expect that, in a well-adapted population, most mutations would be changes for the worse; but it is a big jump to say that all are undesirable because a few, produced in some cases under very peculiar conditions, are weakly. Certainly many of the cases cited in Bateson's *Materials for the Study of Variation* do not suggest an unhealthy individual; and some of the variations, such as supernumerary nipples, can hardly be a handicap. But the weak and strong points of the argument alike must be sought in certain facts which emerge from the study of Mendelian heredity; these we shall notice later.

The next step in the attempt to establish the Lamarckian view is to produce direct evidence of the inheritance of acquired characters. The older instances, such as the experiments of Brown-Séquard, are totally discredited; and it may be added that most of the recent ones require confirmation before they can be accepted without reserve.

Kammerer worked with a particular toad, and with salamanders. Of the latter there are two species to be considered, the first living in damp places, having yellow markings, and bringing forth about twenty young at a birth, which have external gills, and take to the water; the second living in hotter and drier places, black without yellow markings, and bringing forth two or three gill-less young at a birth.

The first experiment was to bring up the black salamander in moist conditions, and the yellow one in hot
dry conditions. The former began to produce larger
families, and the young had gills; the latter had smaller
families, and the external gills were reduced. The
former also took to a larval existence in water.

Kammerer then tried the effect of bringing up yellow
salamanders in black boxes, and black salamanders in
yellow boxes. Before long the black ones developed the
yellow spots, and the spotted ones became black; and
these characteristics were inherited by their offspring.
Even if the yellow offspring of an originally black parent
were transferred to a black box at an early age it was
six months before the blackness began to take effect: at
first the yellow spots merely grew as if the beast was
developing like an ordinary yellow salamander. This
work requires repetition before it can be considered
conclusive. For various reasons most zoologists prefer to
suspend judgment upon it.

Durken has performed a similar experiment on the
Cabbage White butterfly. The pupae are usually greyish
owing to the presence of pigment; but about 4 per cent.
are transparent, letting the green blood show through.
By rearing the caterpillars in boxes with lids of orange
glass, the percentage of green pupae was raised to 67.
The offspring of these, if reared under orange glass, gave
95 per cent. green; and even if reared under ordinary
conditions gave 34 per cent. green instead of the normal
4 per cent., thus showing the inheritance of an acquired
character even though the stimulus was absent.

Some work of Pavlov is of extreme interest and few
observers are more trustworthy than he, but again it
requires confirmation, since parallel experiments on

white rats have shown no trace of the inheritance of a similar acquired character. No scientific experiment can be regarded as conclusive unless it has been confirmed by several independent investigators, in order to eliminate the personal equation, and the possibility of some vitiating factor which has been overlooked. Pavlov rang a bell every time he fed some white mice, as a signal that there would be food in a particular place. Three hundred lessons were needed before all the mice ran at once to that place when the bell sounded. In the second generation only one hundred lessons were needed, in the third only thirty, and in the fourth only five. If this experiment receives confirmation its importance can hardly be exaggerated. But it does not seem to me complete even as it stands. If the mothers ran to the bell when they were pregnant, an environmental factor is introduced. One would prefer to see the experiment made by training male rats only, mating them with untrained females, and then seeing if there was any improvement *at all* in the offspring's quickness of learning. In its class this experiment remains unique; and there is evidence of the total failure to transmit acquired habits in other cases which must be set over against it.

It may further be added that this experiment gives no evidence that eventually a generation will be born which will answer the dinner-bell the first time it hears it, and without the example of other individuals: if such a result were to be established it would indeed be important!

A good deal of work has been done on the effects of alcohol upon various animals, and the results are very discordant.

In the crude method at first adopted alcohol was given with the food. This induced so much dyspepsia that the important aspects were masked by general ill-health; consequently the more refined method of treating with alcohol-vapour was adopted, and found to be quite satisfactory.

Stockard treated guinea-pigs with the vapour to intoxication several times daily. On the whole there was no ill-effect on the parents in any direction, but the offspring were markedly inferior. This would be almost conclusive, but for the fact that after three generations a marked improvement set in!

Pearl treated fowls in the same way, and found that the offspring were definitely improved by the process.

MacDowell found in white rats no structural damage whatever; but a reduction of the capacity of learning, both in the parents treated and in the offspring, whether treated or not, became evident. In fact the effect of the alcohol seemed most marked upon the highest types of nerve-cells.

Hansom found that in white rats (? an inbred strain)[1] there were no ill effects of a physical kind for the first three generations. The fourth was definitely inferior; but the fifth improved.

These contradictory results seem capable of simple explanation, as was pointed out by Pearl in the case of the fowls. If the alcohol-poisoning kills off the weaker germ-cells, the improvement in later generations is easily explained. A more or less immune race may perhaps be established, for some of these experiments have extended over half a dozen years. Of this we shall see that there is other indirect evidence in human beings. Some animals seem to be more affected than others, and one may hazard a guess that a race which is weak at

[1] In a mixed strain one would expect a varying immunity.

the beginning is likely to be more adversely affected. Finally the evidence suggests, though it does not prove, Horsley's contention that the effect of alcohol upon the higher levels of the brain is deleterious, at any rate when large quantities are absorbed.

The necessity for caution in coming to a decision is shown by the experiments of Bentley and Griffith on rats kept in cages rotating at one to two revolutions per second. The inhabitants soon got used to the motion, but on removal from the cages showed the kind of disequilibration which we feel slightly on landing after a rough voyage. Their eyes rolled, and they staggered; and the effects were correlated with the direction of rotation of the cage. Their offspring inherited the disequilibration in many cases.

Detlefsen however showed that rats which had never been rotated might exhibit the same symptoms if they were suffering from an infection of the labyrinth of the ear; that the offspring contracted the same disease; and that according to the side which was affected you got left- or right-handed disequilibration. Since, in the original experiment, many cases of a discharge from the ear were noticed, it is reasonable to assume that here we have a case of infection, and not one of the inheritance of an acquired character. Rotation might produce a predisposition to the disease, and infection would do the rest. It must however be admitted that if future experiment confirms the suggestion that the offspring show the same direction of disturbance as the parents (right- or left-handed), there may be more in the case than this.

It will be seen, then, that though the inheritance of acquired characters is undoubtedly suggested by some

of the broad features of evolution—such as the development of the *Trochophore* larva into the worms and
molluscs, the adaptation of the flat-fishes, and the
general probability that an evolution which is in the
main adaptive should make direct use of the adaptation
of the individual—yet the direct experimental evidence
is very far from conclusive. The cases most thoroughly
investigated have so far proved capable of another explanation, and a more likely one. Infection, or the
killing-off of weak or ill-adapted germ-cells, seems to
account for the phenomena. The more striking cases
still await verification.

Meantime the fundamental difficulty, that of imagin-
ing any suitable mechanism, remains. No one questions
the fact that a given environment may strengthen or
weaken the germ-cells, and so, by selection in the widest
sense, give rise to new types; but it is extraordinarily
difficult even to conceive a mechanism which could
affect the germ-cells in such a manner that some
adaptive modification in a group of cells could reappear
in the resulting individual in the particular group of
corresponding cells after the thousands of cell-divisions
necessary for the production of a body from a fertilised
ovum had taken place—that these cells and no others
should develop the character. Difficult, but perhaps not
impossible.

We have a little evidence which may perhaps point in
this direction; and which is at least enough to make us
pause before saying that there is no reason at all for
believing that such a process, in some more generalised
form, is even possible. We have already mentioned the
hormones; those chemical mechanisms which, secreted
by a specific group of cells in the body, are poured into

the blood and stimulate the development of some particular organ at a distance. A good example is the secretion of the pituitary body, which, with others, is responsible for the development of the secondary sexual characteristics.

Evidence is accumulating that, from the earliest stages of segmentation of the egg, the building up of the body is controlled in this way, starting from the stage of differentiation of the organs. Such evidence emerges as soon as recent work on regeneration of lost parts, and on artificial interference with developing embryos, is considered in detail. It can hardly be dealt with here, since it can only be appreciated after fairly wide study. But it is clear that if the development of an organ is conditioned by the secretion of a hormone, and there is no evidence that in most cases the hormone ceases to be secreted in the young adult, there must also be another type of messenger whose action is antagonistic to the first; and that finally a balance is struck between the hormones and these anti-bodies, which are called chalones.

Of the chalones comparatively little is known, but an experiment of Guyer and Smith is of sufficient importance to demand mention.

The lens of the eye of a rabbit was pulped, and the extract injected into a fowl. In the blood of the fowl a chalone or anti-body was generated; and when this blood was injected into a pregnant rabbit, the lenses of the eyes in the young did not develop properly. Moreover the offspring of these rabbits, though untreated, were also defective; and the defect was transmitted by the male as well as by the female.

Darwin's imaginary "pangens" were purely theo-

retical constructions: here we have a chemical body
performing exactly the kind of function which he attributed to them hypothetically. The importance of this
cannot be over-estimated. If other similar cases can be
demonstrated we have something approaching to the
required mechanism for the transmission of an acquired
character; though it is still a far cry to a full Lamarckian
position; and it may fairly be said that proof of the
transmission of acquired characters in that sense is
almost as distant as before. A hormone and chalone
which will ensure the transmission of the mental characters produced by desirable education and the sup-
pression of those produced by undesirable education are
still unimaginable. And it is practically impossible to
conceive of a hormone basis even for the salivary-reflex
and metronome association experiment of Pavlov,[1] yet
it is an exactly similar experiment that Pavlov depends
upon—the dinner-bell and mouse—for proof of the
inheritance of acquired characters. We must not forget
that the dog, trained to expect food when the metronome
beat 100, and not to expect it when the metronome
beat 50, developed neurasthenia when it beat 75. To
attribute this to hormone reactions is to stretch probability almost to breaking-point; even though we may

[1] The reference is to Pavlov's work on conditioned reflexes.
Briefly, dogs "learned" to salivate at the sound of a metronome
beating at a definite rate, associated with the feeding-time, instead
of at the sight of food. One stage in the chain of events is suppressed,
and the associated salivation is moved a step forward. Pavlov is
endeavouring to prove that all education of the mental processes
is capable of a mechanistic explanation. My point here is that even
if we accept this, the simplest of mechanistic interpretations (which
I do not), the difficulty of a Lamarckian explanation of the
inheritance of education is startlingly emphasised.

agree that neurasthenia may possibly be associated with, or conditioned by, hormone unbalance, psychological conflict is here the obvious cause.

Quite possibly, a muscularly active person might produce a lot of muscle-developing hormone, and even pass it on to his descendants, though this remains to be proved; but it is difficult to believe that a blacksmith might produce a hormone which develops a horny epithelium in a few special places, and hand that down to his children, so that a horny layer should develop in exactly the same places. Added to which, a blacksmith's children do actually have hands just as soft as those of any other babies!

NOTE. We have observed that the evidence from the botanical side has hardly received the attention it deserves. Whatever view may finally be adopted concerning this evidence, it is important to remember that in plants the sex-cells are differentiated far later than they are in animals. This clearly increases the possibility of somatic influences affecting the next generation.

Chapter IV

HEREDITY—THE PRINCIPLES OF MENDELIAN INHERITANCE

From every side our studies of evolution have converged upon heredity. It is time to set down some of the elements of that subject.

There are only two possible routes of approach, the first by statistics, the second by the detailed examination of individual cases.

Of the statistical method I propose to say little. It involves difficult mathematical analysis beyond my powers; and it cannot help us far along the particular road we are traversing.

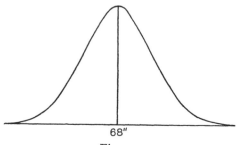

68"

Fig. 27.

It is easy enough to graph any measurable feature in any organism; and if your mathematics be adequate it is possible to get a good deal of knowledge from the graphs. We need only consider one or two of the easiest examples of the method.

If human stature be graphed, height being taken as the abscissa and number of individuals of that particular height as the ordinate (it is better to confine the graph to individuals of one class and sex, so that accidental differences due to nutrition, lack of sunlight, and such things may be eliminated, since they are apt to complicate the issues), it will be found that the graph has a very definite and symmetrical shape, as fig. 27 indicates.

This means that the great majority of the members of the population have a height of, let us say, 68 inches.

There are a good many who differ from this by an excess or defect of an inch, fewer by an excess or defect of 2 inches, and very few indeed by an excess or defect of 6 inches. The greater the number of individuals examined, the nearer does the graph approach to ideal smoothness and symmetry.

Now if two pennies are tossed a great number of times, say 1000, it will be found that about 500 times they turn up one head one tail; 250 times they turn up two heads, and 250 times they turn up two tails. That is to say the probability of their turning up is as

$$1\,HH : 2\,HT : 1\,TT.$$

If three coins are tossed the probability becomes

$$1\,HHH : 3\,HHT : 3\,HTT : 1\,TTT.$$

Evidently what we are doing is expanding the expressions $(H + T)^2$ and $(H + T)^3$.

If the power is high—that is to say, if the index is large—we find that we get a smooth curve. Thus even with a power of 10 the curve approaches the ideal pretty closely. The curve is of course obtained by plotting the arrangement of heads and tails as abscissae, and the number of instances as ordinates.

In general, the same result is arrived at, whether we actually toss the coins a sufficient number of times, or expand the series $(H + T)^n$; and the equation to the curve is $y = e^{-x^2}$. This illustrates the fact that if you take a large number of factors, which tend to operate equally in either direction, the results when graphed will fall along such a curve, which is named the "curve of normal probability".

Human height, which depends upon the dimensions of a large number of skeletal elements, is admirably suited to produce such a result. Galton examined the marks of the Mathematical Tripos at Cambridge, and found that again the same curve was obtained. Any examiner, in fact, who discovers that his marks for a paper do not approximately give such a curve can be certain of one of two things: either he has marked the paper badly, or the paper itself was an unfair one.

The perpendicular drawn to the highest part of the curve is called the "norm", and the curve is usually symmetrical about it; though this last is not always the case. Thus, if we graphed the number of children per marriage at the beginning of last century, when large families were frequent, we might find that the average number was five, but that the curve stretched away to the right into the twenties. Clearly such a curve could not be symmetrical, since it is impossible to have less than no children. Skew-curves are, then, found in some instances.

Now when the lengths of the seed-pods of three closely allied species of *Oenothera* were measured by de Vries it was found that, while two gave roughly the curve of normal probability, though with different norms, the third showed two maxima. Evidently the third species was tending to split up into two different races. Similarly Miss Bateson showed that the earwigs of the Farne

Islands are tending to split into two races, one of which has long pincers at its hind end, the other the usual short ones. The graph shows this clearly.

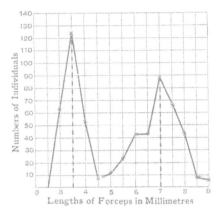

Fig. 28.

Many other things can be learned from graphs of this kind, but we cannot discuss them here.

Another type of graph gives valuable information of a different kind. Suppose the heights of fathers are plotted as ordinates against the heights of their sons plotted as abscissae, we are bringing our statistics definitely into the region of heredity.

Let us say that the mean height of the population of fathers is 68 inches. If there were no correlation between the heights of sons and fathers, the sons' heights would lie roughly along the horizontal line passing through 68 inches: their average would be 68 inches. If sons were exactly the same height as their fathers, the line representing their heights would lie at an angle of 45 degrees to the horizontal. Actually it lies between these two extremes, and the tangent of the angle of slope is known as the "coefficient of correlation". The graph shows that, while tall parents will tend to have tall sons, those sons will be less tall than the parents. This is generally expressed by saying that there is regression towards the mean of the population.

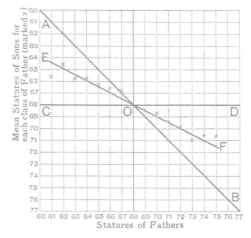

Fig. 29.

Since, in fact, the mothers' heights come in as well, it is desirable to use what Galton called the "mid-parent" for the ordinates—an abstraction which takes the mean of the fathers' heights and the mothers' heights, allowing for the fact that women are normally shorter than men.

The numerical value of the correlation-coefficient is in the neighbourhood of 0·5 as a rule: that is to say, the angle of slope is somewhere about 27 degrees. Though it varies appreciably when different characters are graphed, it is generally in this neighbourhood for continuously varying qualities, even though these may be not physical, but mental qualities such as conscientiousness or ability or vivacity.

Evidently such statistics can teach us a good deal about the general lines along which inheritance goes, though they can tell us nothing whatever about individual cases.

Karl Pearson, working along a different line, has arrived at another result of extreme interest to the general theory of evolution. Suppose that you wanted to increase some characteristic of the population by an amount which exceeded the norm of the population by a quantity a. You would only allow those individuals which exceeded by a to breed. F_1 (the first filial generation) will show $0.62a$; F_2 will show $0.82a$; F_3 $0.89a$; F_n $0.92a$. If this be graphed, it will be seen that we obtain an asymptotic curve, analogous at least to the curve we should get by attempting to establish a pure line by selection.[1]

If selection be stopped after the nth generation, and the stock be promiscuously inbred, Pearson calculated that the F_1 generation from this point would show $0.86a$; F_2 $0.81a$; F_3 $0.77a$; and F_{100} $0.51a$. Thus there is a gradual regression of the race towards the mean of the population. This suggests that it is impossible to establish a race by selection, at any rate where continuous variations are concerned; and this is in agreement with what we noticed in our study of Johannsen's work on pure lines.

[1] The time a horse takes to cover a mile in American trotting races gives another pretty example of an asymptotic curve obtained by selective breeding. This curve shows two sudden rises, however —one due to the introduction of ball-bearings in the wheels, the other to the introduction of a moving wind-screen in front of the horse and vehicle!

In another matter the evidence of statistical methods is of more doubtful value.

Galton, founding his work on the biological conception of the fertilisation of an ovum by a spermatozoon, arrived at what is known as the "Law of Ancestral Heredity". This states that, of the characters of an individual, on the average one-half are derived from the parents; one-quarter from the grandparents; one-eighth from the great-grandparents, and so on. As a statistical statement of an average this may possibly be true; as a statement of biological fact it is woefully wide of the mark.

Karl Pearson has since removed the law still farther from biology into the region of pure mathematics, and has arrived at a series which diminishes more rapidly. Again, his conclusion may be true statistically; but neither form of the Law of Ancestral Heredity has any great significance for the biologist. He has other irons in the fire.

As early as 1760 Kölreuter carried out a series of experiments on plant-hybridisation. It must be remembered that at this date it was not known that fertilisation consists in the union of one male gamete with the ovum; nor was the importance of the nucleus even guessed. Kölreuter very nearly discovered that one pollen-grain fertilises one ovule; and he definitely discovered the fact that reciprocal crosses were identical—that is to say, that you got the same result whether you fertilise the ovule of A by the pollen of B, or the ovule of B by the pollen of A. This discovery was extraordinarily important, for the part played by the male germ was not then known, nor was it suspected that it was as essential as that of the female. Indeed, speculation about fertilisation had got no farther than the idea that it was brought about by the mixing of two fluids.

Kölreuter also found that reciprocal crosses were not always equally easy to achieve: sometimes the ovule of A could be fertilised by the pollen of B, but not *vice versa*. He also noticed both the vigour and the sterility of hybrids; indeed this last observation was the starting-point of the unsatisfactory definition of species to which we have already alluded. It is true that crosses between distant species are generally sterile, but this is by no means wholly true of species which are closely related.

Finally Kölreuter re-obtained the parental type again by crossing and recrossing the hybrid with one of the parents.

About the same time Thomas Andrew Knight, who discovered tropisms a hundred years before biologists had begun to consider them at all, and whose experimental work was of the highest rank, drew further attention to the fact that a hybrid was apt to be far more vigorous than its parents, even though it was sterile. The hardiness of a mule is proverbial; and I have only heard of one mule—or jennet, I do not know which—that was not sterile; and as the information came to me indirectly from a Greek of Salonica who was selling remounts for the Serbian army, it can hardly be considered as unimpeachable evidence.

Knight's main object was to see whether debilitated inbred stocks could get a new lease of life from crossing with a slightly different race; but in the true scientific spirit he experimented on a far wider basis. Thus of peas he writes, "By introducing the farina of the largest and most luxuriant kinds into the blossoms of the most diminutive, and by reversing the process, I found that the powers of the male and female in their effects on the offspring are exactly equal. The vigour of growth, the size of the seeds produced, and the season of maturity, were the same, though the one was a very early, and the other a very late variety. I had, in this experiment, a striking instance of the stimulating effects of crossing the breeds; for the smallest variety, whose height rarely exceeded two feet, was increased to six feet, whilst the height of the large and luxuriant kind was very little diminished".

With this may be compared Mendel's statement: "The longer of the two parental stems is usually exceeded by the hybrid, a fact which is possibly only attributable to the greater luxuriance which appears in all parts of the plants when stems of very different lengths are crossed".

When we realise that a hybrid between a radish and a cabbage reaches 10 feet or more in height, with a corresponding spread, the truth of Knight's observation is firmly brought home.

Of the reason for this vigorous growth and hardiness in hybrids we shall speak later: for the moment it is sufficient to notice the fact, and to draw attention to the partly fallacious law that was founded upon it.

The not unnatural inference from such observations was that the efficient cause was fertilisation by a race well removed from that of the female: that exogamy itself was what mattered.

This conception led to the statement that self-fertilisation, or close inbreeding, must alike bring degeneration. Such an assertion is incorrect, as we shall see. A fair number of plants at any rate are always propagated by self-fertilisation, and they do not seem to suffer; though in the vast majority of cases cross-fertilisation is the rule. The point of view became crystallised in Darwin's aphorism: "Nature abhors perpetual self-fertilisation"; and it is only the work of quite recent years which has exploded the fallacy. Clearly, it is a fact that cross-fertilisation is usually advantageous, as is shown both by the care taken to insure it, and the experience of breeders of the dangers of too close inbreeding. The fallacy lies in the assumption of some magical significance in cross-fertilisation *per se*. We shall find that there is a perfectly simple explanation of the observations. But the so-called "Knight-Darwin Law of Intercrossing" long dominated the situation.

Other workers, especially Naudin, came very near to making discoveries of far-reaching importance, but one and all missed the goal through not seeing the importance of exact numerical records of their observations upon hybridisation. It was left for Gregor Mendel, Abbot of the Augustinian monastery at Brünn, in Austrian Silesia, to make the great advance. Yet, by the irony of fate, his work remained buried in the Report of a local Natural History Society for nearly forty years; and by this mischance biological science stood for forty years before a locked door whose key lay close at hand.

In the first year of the twentieth century de Vries, Correns, and Bateson drew attention to the long-lost work.

It is fair to say that Bateson, a zoologist possessed of a mind perhaps unequalled in its sphere since Darwin, realised more completely and more immediately than anyone else the epoch-making nature of Mendel's work. He alone seemed fully to understand that it opened up entirely new vistas in the theory of evolution; and he was not afraid to fight for what he knew to be true.

Those who attended his first course of lectures on the subject at Cambridge, among whom I was fortunate enough to be numbered, can never forget the impression. They may not have understood altogether; for it was a new idea, and one was so engaged in trying to grasp the facts that one had little hope of gaining a deeper insight; moreover Bateson was so complete in his own mastery, and so full of his anticipations, that I at least found him hard to follow at times. All the old work on hybridisation—half-races and false hybrids, atavism, biometry—it all poured forth, and here was the clue to everything. But one did realise that one was seeing the birth of a new order in biological thinking; and to visit his house and see the actual experiments in progress was a never-to-be-forgotten experience.

Bateson at once took up the gage that opposing scientists threw down; and the fight was a sharp one. In the preface to his first book on the subject he begins with the statement that evolutionary theory had simply got stuck in the jungle. "Such was our state when two years ago it was suddenly discovered that an unknown man, Gregor Johann Mendel, had, alone and unheeded, broken off from the rest—in the moment that Darwin was at work—and cut a way through. This is no mere metaphor, it is simple fact. Each of us who looks at his own patch of work sees Mendel's clue running through it: whither that clue will lead, we dare not yet surmise."

Since Bateson wrote those words the clue has been followed far; though not yet to the end. It is our next business to explain first the nature of the clue, and then to look at a much-simplified map of the way along which it has led us; and even to look a little at the jungle ahead, as it were from a tree-top, and guess whither we are tending.

For brevity and simplicity our method will be to follow step by step the developments and modifications of Mendel's theory in their very barest form, leaving difficult complications on one side, and illustrating each point by one example and one only; but it must be understood at the outset that these conclusions do not rest upon single instances, but rather, that every stage has been tested and retested again and again by other experiments.

We shall, however, go far more thoroughly into this than into other matters, even at the risk of over-weighting one part of our work, for it is of fundamental importance and from this stage onwards Mendel's work will never be far from our minds. Later we shall try to show how it has altered our views about the theory of evolution.

Mendel began his work on peas. These plants are particularly suitable since it is easy to ensure self-fertilisation, or cross-fertilisation, with any desired pollen. The flowers are large, and easily protected from accidental pollination by paper or muslin bags.

One of Mendel's experiments will serve to give us the general idea; but only reference to his own paper can give any idea of the meticulous care he took to eliminate disturbing factors.

A tall pea was crossed with a dwarf pea. In the first generation, which we may call F_1, all the peas were tall. The individuals of this generation, when either crossed with each other or self-fertilised, gave, when large numbers were taken, 25 per cent. short, and 75 per cent. tall. If this F_2 generation was self-fertilised, it was found that whereas the short peas bred absolutely true to shortness for an indefinite number of generations, never throwing another tall, the tall ones behaved differently. One-third of them bred true indefinitely, while the remaining two-thirds behaved exactly like those of the F_1 generation, giving 75 per cent. tall to 25 per cent. short; and these behaved in their turn in the same manner.

Thus it is clear that in the F_2 generation, while the shorts contain no element of tallness, one-third of the talls contain no element of shortness, but the remaining two-thirds of the talls contain shortness. Again, it is clear that tallness can mask shortness, but shortness cannot mask tallness. A simple scheme will make the results obvious:

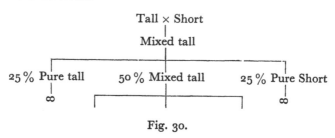

Fig. 30.

Mendel's explanation was simple, and in the main patently true. He suggested that such characters as tallness and shortness formed a pair of what he named "allelomorphs". The plant could carry both tallness and shortness, but the germ-cells were "pure" as regards the allelomorphs: that is to say, they could carry one or the other, but not both. Thus in a pure-bred pea the gametes were all of one kind, but in the hybrid the gametes were of two kinds. The first had all "tall" or all "short" ova, and all "tall" or all "short" pollen-nuclei. The second had "tall" ova and "short" ova, "tall" pollen and "short" pollen, two kinds of each gamete, and their chance of meeting is resolved into a mere question of simple mathematics. It is heads and tails for two pennies again.

The probabilities are 1 **TT** : 2 **TS** : 1 **SS**; and the ratio will be the more closely followed, the greater the numbers used.

The other point jumps to the eye. It is clear that where tallness and shortness co-exist in the same plant, the tallness masks the shortness. In other words, there are two kinds of tall plants, one pure, the other hybrid, but practically indistinguishable except by breeding them; while there is only one kind of short plant.

Thus, all the first hybrid generation appear to be tall, while the second generation gives 25 per cent. short, and 75 per cent. tall; but further investigation shows that really we have 25 per cent. pure tall, 50 per cent. hybrid tall, and 25 per cent. pure short. Mendel called tallness the "dominant" character, shortness the "recessive".

We may now rewrite our scheme, making it clearer by using the conception, confirmed both by breeding experiments and by observation of the nuclear phenomena, that the pure-bred plants have a double dose of the character in question, the hybrids a single dose of each of the allelomorphs. The scheme then becomes:

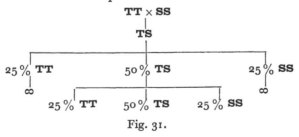

Fig. 31.

Mendel's great discovery, then, was the purity of the gametes for a given pair of characters.

That dominance is not by any means a universal phenomenon may be shown by a more recent experiment.

There is a breed of fowls called Blue Andalusian, characterised by smoke-grey colour with beautiful black pencillings round each feather. Such birds refused to breed true. They always threw a certain number of black and of splashed white offspring, which were religiously discarded. When numerical observations were made it was found that there were about 50 per cent. of the Blue Andalusian chicks, and 25 per cent. of each of the other types. Further, it was found that if the blacks were crossed with the splashed whites, all the chicks were of the Blue Andalusian type; but these behaved no better than their predecessors in the matter of breeding.

The explanation is obvious. The Blue Andalusian is not a pure breed; but neither the black nor the splashed white exhibit dominance.

In the convenient phraseology of the Mendelian theory, the Blue is a "heterozygous" form, while the black and the white are "homozygous" in regard to colour. Again a scheme will make this clear:

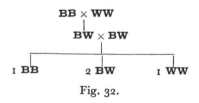

Fig. 32.

where **BW** is the heterozygous Blue Andalusian.

The failure of dominance in some hybrids is a fact of vital importance, as will appear later.

We may now consider the Mendelian ratio when there are two pairs of characters involved; for Mendel worked out the theoretical numbers for all possible cases, and tested his conclusions in the simpler cases.

Here we will anticipate the next modification of the original idea by using different symbols. Suppose a tall purple sweet pea is crossed with a short white one. Tall is dominant to short, purple to white. We will express the tall purple plant by the symbols **PPTT**, implying a double dose of tallness and purpleness. For the short white plant we will write **pptt**, the small letters simply indicating the absence of the dominant factors, purpleness and tallness.

The first generation will all be purple tall, but the plants will be heterozygous for both purpleness and tallness.

The second generation shows, when self-fertilised, four different kinds of plants, purple tall, purple short, white tall, and white short. How this happens is clear when we think of the gametes that the F_1 hybrids will produce. **T** and **t** cannot go into the same gamete, neither can **P** and **p**, since the gametes are pure for each of the allelomorphic pairs; but all other combinations are possible. Inspection shows that there will therefore be four kinds of pollen-nuclei, and four kinds of ova. These will be represented by the symbols **PT**, **Pt**, **pT**, and **pt** in each case. To see how these will combine, the simplest plan is to construct a chess board of sixteen squares. Starting from the top left-hand corner, the symbols for the gametic constitution of the ova are written down in order, each repeated four times *horizontally*. The symbols for the pollen-nuclei are now written in the same order in the squares, beginning at the same place, and repeating each four times *vertically*. A diagram will make this clear.

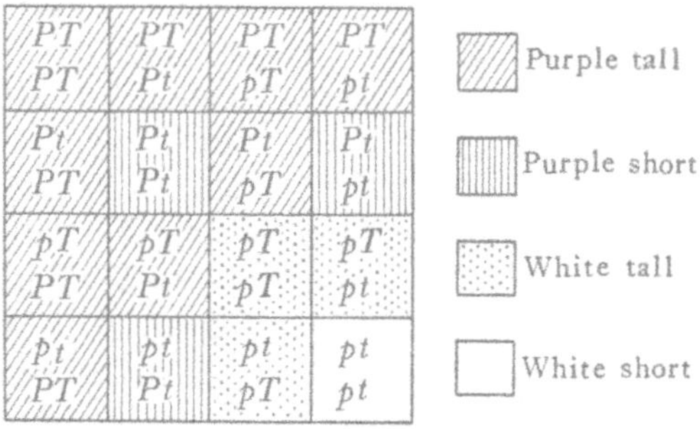

▨	Purple tall
▥	Purple short
▨ (dotted)	White tall
▢	White short

Fig. 33.

If we now examine our chessboard we shall find that there are nine squares which contain both **P** and **T**; though only one will be homozygous for both characters, the others being heterozygous for one or both; these will all appear as purple talls. There will be three squares which contain a double or single dose of purple with no tallness; these will be purple shorts. There will be three which contain tallness but no purple; these will be tall whites. And finally there will be one square containing neither purple nor tall, which must be a short white. Thus in F_2 there will only be one pure dominant and one pure recessive in every sixteen plants, where large numbers are taken. This $9:3:3:1$ is known as the "dihybrid ratio". It may appear as $9:3:4$, or $9:7$ when some of the types are indistinguishable, as we shall see. For the tri-hybrid ratio sixty-four squares would be required, but we need not enter into that, for it introduces nothing but mere technical complications.

Very soon it became clear that it was possible to improve upon Mendel's statement of his discovery. In particular, his idea of allelomorphic pairs had quickly to be abandoned in favour of the simpler and more satisfactory conception that we were concerned only with the presence or absence of a given character. An illustration—one of the first to suggest the need of this modification—will make the point clear.

Fowls have combs of various types. One, known as the Pea comb, has three rows of blunt tubercles running lengthwise; another, the Rose comb, is somewhat triangular, projecting backwards, and covered with small papillae. Both of these are dominant to the common Single comb (fig. 34). The question was, what would happen when you crossed two dominants? The expectation was that one would prove dominant to the other, in the simple Mendelian way. But the unexpected always happens. Rose was crossed with Pea. What would result? The answer was a Walnut!

The Walnut comb is not unlike half a walnut coming down over the top of the beak, and is characterised by a number of bristles.

But the F_2 generation produced another surprise; for besides Walnut, Rose, and Pea combs there appeared a few Single combs. (Self-fertilisation being impossible, of course the members of the F_1 generation had to be crossed with each other.)

When sufficient numbers had been crossed it became clear that the ratio was 9 Walnut to 3 Rose to 3 Pea to 1 Single.

Evidently the Single character must have been a recessive which was masked by the dominants Rose and Pea, being present in one or other, or in both.

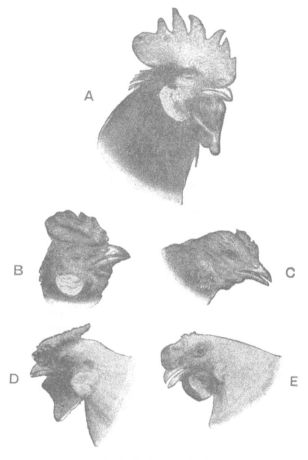

Fig. 34. Types of combs in fowls: A, single comb (cock); B, pea comb (cock); C, pea comb (hen); D, rose comb (cock); E, walnut comb (young cock). (From Bateson.)

On this assumption we may write the pure Rose as **RRppSS**, and the pure Pea as **rrPPSS**. F_1 will then give **RrPpSS**, assuming that there was a double dose of Single in each. The gametes are **RPS, rPS, RpS** and **rpS**; and the chessboard will work out thus:

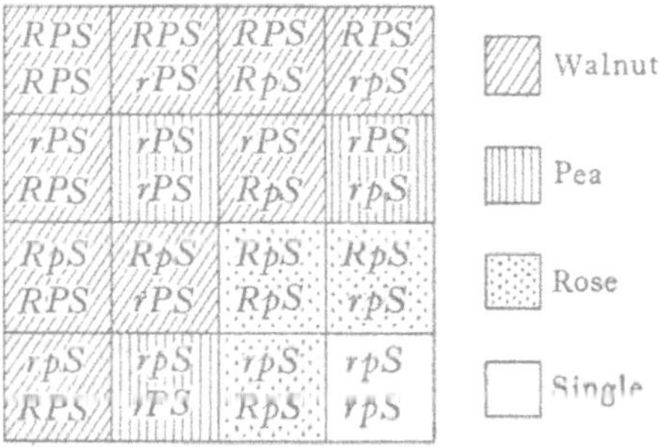

Fig. 35.

So far, so good; but we must not be content simply to assume that there is Single in each comb: the hypothesis must be proved.

There is a combless fowl called the Breda, in which two tiny knobs of flesh take the place of a comb. When crossed with a Single in a preliminary experiment, the F_1 generation was all duplex single. The element of duplicity came obviously from the Breda; and it was equally clear that the character of Single comb did not come from the Breda, since it emerged at once when it was introduced. The Rose comb was next crossed with the Breda. The F_2 generation was a little complicated, since the element of duplicity had also been introduced, but the expected result occurred—that a few of the fowls had Single combs, either duplex or simplex. The Single could only have come from the Rose, since the Breda does not contain it. With the Pea comb a similar result was obtained. The proof was complete. But it is very difficult to explain this case on the Mendelian hypothesis of allelomorphic pairs. Pure Rose and pure Pea can be crossed with pure Single, which then behaves as an ordinary recessive. Yet, when Rose is crossed with Pea, in the F_2 generation, Singles emerge, though no Single was put in.

The explanation adopted, which has ousted the allelomorphic one in the general expression of Mendelian ideas, is what is known as the "Presence and Absence Theory". In this particular kind of case, where dominance exists, the dominance is supposed to be due to a factor which is absent in the recessive. In general, the conception of alternative characters (allelomorphs) has been replaced by that of the presence or absence of a factor. Thus a sweet-pea is short because it lacks the character of tallness.

It will now be clear why we gave up the original shorthand form which used **T** and **S** as symbols, replacing it by **T** for tallness and **t** for absence of tallness. A further convenience of this notation is, that the homozygous and heterozygous types are still distinguished at a glance. Thus **TT** is homozygous for tallness, **Tt** heterozygous; while a plant can only be short if it is homozygous **tt**. This simplified conception constitutes a real advance in the theory.

Again, Mendel worked with pairs of characters which were not likely to interact. But the factors which govern such things as coat-colour in mice are very likely to interact, and their investigation opened up a new chapter.

The *locus classicus* for the start of this work was an

unexpected discovery of Cuénot's in working out the Mendelian ratios in mice. I have good reason for remembering it, for in Part II of the Tripos there was in my day a cruel barbarity. At the end of a seven-hour Practical examination, which came last of the papers in a particularly hot June, I was called to undergo a long *viva voce*. I believe I acquitted myself respectably among most of the exhibits that had to be discussed, though I cannot be sure, as by that time I was a mere automaton; but finally came a cage of mixed rats, whose lineage had to be suggested. These were white, black and ordinary; and one of the white ones yawned steadily throughout the interview. I knew that albinism was a recessive character, and I had not heard of Cuénot's quite recent work. So I firmly asserted that whatever crosses were made with the wild or the black types, the white would behave as a simple recessive. I could not think, and the rat provoked in me an almost irresistible desire to yawn too.

Since that day I have never forgotten Cuénot's experiment!

The wild type of coat-colour is known as "agouti". White is recessive to any colour. Therefore the expectation for agouti crossed with white is F_1 all agouti; F_2, 3 agouti : 1 white. This normally occurs, but Cuénot found in certain cases that the F_2 ratio was 9 agouti : 3 black : 4 white. Where did the blacks come from?

Clearly we are dealing with a dihybrid ratio with the last two terms indistinguishable; and since white is the mere absence of colour, there must be another factor involved.

Further experiments led to the following conclusion. Agouti is black with a dominant greying factor added.

For any colour to develop it is necessary, however, that a colour-developing factor should be present in addition. The supposed white was really a black which had no colour-developing factor, and consequently appeared white. In this case black *is* white!

We may denote our homozygous agouti as **CCGGBB**, and our albino as **ccggBB**, both parents being homozygous for black. The gametes will be of four types, **CGB, cGB, CgB,** and **cgB**. A chessboard will make the rest clear.

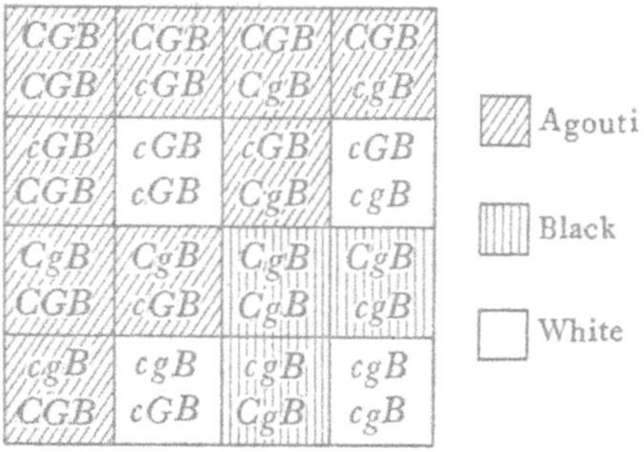

Fig. 36.

Anything containing both **C** and **G** will be agouti, **C** without **G** will be black, and anything without **C** will be white, whether it contains **G** and **B**, or only **B**. The development of a character is here seen to depend on the interaction of two factors; and the discovery of this fact paved the way for important advances.

Of course, the observation had to be confirmed, both by many experiments designed to test the hypothesis in the particular cases, and by experiments of a like kind in both plants and animals. Such tests were successful.

The next point to arise was of remarkable interest,

and it led up to Bateson's famous address to the British Association just before the War.

Long ago Darwin noticed that when certain pigeons were crossed, eventually a few individuals appeared which closely resembled the ancestral Blue Rock. He made use of this case of atavism in arguing that domesticated breeds had been produced from wild species by a process of intelligent selection. The experiment was repeated in a simpler form, and worked out in Mendelian terms. Black Barb was crossed with White

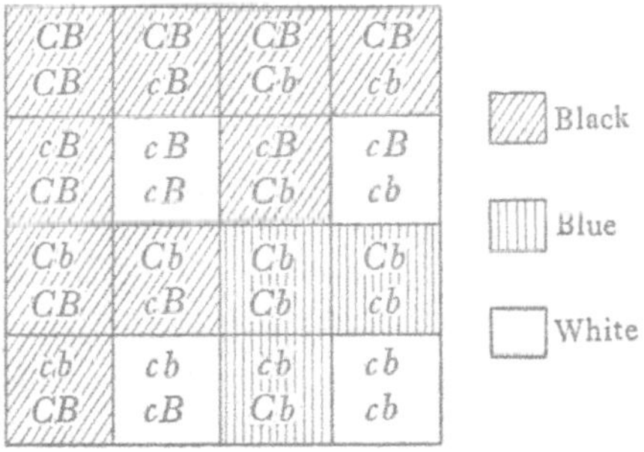

Fig. 37.

Fantail. F_1 was black, white-splashed; F_2 gave 9 black : 3 blue : 4 white, some of the black and the blue birds being white-splashed.

Neglecting this complication we can say that the Black Barb had the essential constitution **CCBB**, the Fantails **ccbb**, where **C** is a colour-factor (blue), in whose absence no colour can develop, and **B** is the colour-modifier, in this case black. The gametes are **CB, cB, Cb,** and **cb,** and the chessboard gives the ratio 9 : 3 : 4.

The real interest of the case lies in its showing that we get reversion to the ancestral type, by a rearrangement of the Mendelian factors. In this particular instance the process happens to consist in the elimination of a modifying factor. In the agouti mouse it was the re-combination of two factors which had got separated in breeding the black (or the albino-black, and the albino-grey) and this seems to be the more usual case.

Upon such considerations was founded Bateson's famous attempt, outlined in his British Association Address, to formulate a theory of evolution based mainly upon the Mendelian conception. In its extreme form, which would suggest that evolution was brought about by the separation of factors originally combined, one would have to look upon Shakespeare as an amoeba with a good deal left out—a view which will hardly commend itself to the philosophically-minded biologist. Indeed the less extreme form of the theory, which allows for the production of new types by the combination of factors hitherto uncombined, is really faced with the same sort of difficulty, even if it allows for new factors arising in ways other than those recognised by recent Mendelian advances, such as the breaking of linkages (*vide infra*); and, if it does admit such unrelated happenings, it gets away at once from the mathematically precise theory which seemed so promising.[1]

The fact is, that it grows less and less probable that we shall find one single cause to account for evolution;

---

[1] As a matter of fact it is becoming increasingly clear that, in man at any rate, the Mendelian ratios do constitute only a first approximation to the truth. It looks as if there were other things to be reckoned with as well. Therefore I make no apology for referring in several places to the probable influence of hormones, whose production *may* perhaps be determined by other genes, or may not, even though this work may seem less certain and clear-cut than that which goes before.

and it is a wise policy to keep an open mind. Evidence is slowly accumulating.

There is no doubt that the Mendelian factors have played an important part. There is no doubt that natural selection has played an important part. The first supplies a large range of variations; the second is at least one great determinant of their survival value. Another determinant is the general compatibility of the genes[1] (Mendelian factors), as we shall see. But it is also possible, though in my judgment as yet unproved, that the Lamarckian factor has played its part too, perhaps mainly through the hormones; the *a priori* probability is considerable, though the proof is incomplete, and the method hard to conceive.

So far, then, we have examined the simple Mendelian theory, and the numerical ratios for one and two pairs of characters. We have seen that dominance is by no means a universal feature: indeed every stage from complete dominance to no dominance at all is found. We have seen why it is necessary to replace the conception of allelomorphic pairs by the Presence and Absence Theory. We have seen that where factors interact there may appear to be a disturbance of the ratios, and unexpected characters may emerge; but that closer investigation reveals these cases as really confirming the theory. We have seen how the appearance of atavistic forms is explained.

The general importance of Mendelian theory is so obvious that we need not dwell on it. It gives a method of analysing hereditary characters, and of combining desirable ones. A recessive can be established in two

[1] "A word presumably derived from the Greek" as Punnett caustically remarks in a footnote.

generations, a dominant in three (the need of the extra generation in this case is due to the fact that it is frequently impossible to distinguish a homozygous dominant from a heterozygous, except by breeding). The combination of desirable characters opens up great possibilities. Biffen's Wheat is the classical example. Professor Biffen managed to combine the rust-proof quality of a wheat which bore but poorly with the heavy-bearing of another wheat, and with the desirable milling qualities of a third. (Rust is a fungus-disease due to *Puccinia*, which causes great loss by diminishing the vitality and the yield of the plants.) Further, real understanding of the facts of inbreeding and out-breeding, which we are about to consider, is bound to have a great effect on both stock farming, dog and horse breeding, and on plant culture, as it becomes more widespread.

Chapter V

THE MECHANISM OF MENDELIAN INHERITANCE. SEX. INBREEDING AND OUTBREEDING

At this stage it will be well to notice how perfect is the mechanism for Mendelian segregation which has been discovered by the cytological investigation of nuclear division. In normal cell-division the chromatic thread is first differentiated from the chromatic material of the network and nucleolus, and then breaks up into a specific number of chromosomes—anything from two to a hundred or more. Then the chromosomes divide longitudinally, one-half going to each pole of the spindle. But more than that, not only each chromosome, but probably every granule or chromomere in that chromosome likewise splits. If then we conceive of the chromomeres as the physical base of either one gene or of a small group of genes, it is clear that care is taken to supply every cell with every gene. (Perhaps this statement is an exaggeration: it seems probable that cells of a differentiated tissue lack some elements present in the more generalised cells of the first stages of cleavage; but this point need not concern us, and we know little about it.)

In the formation of the gametes there is a vital difference, which leads to the halving of the number of chromosomes. During the first division the chromosomes fuse in pairs, either ring-wise or by some other method such as tetrad-formation; and then separate, whole chromosomes (to which something has however happened) going to the poles of the first spindle. Without a resting-stage the second spindle forms, the chromosomes split and the

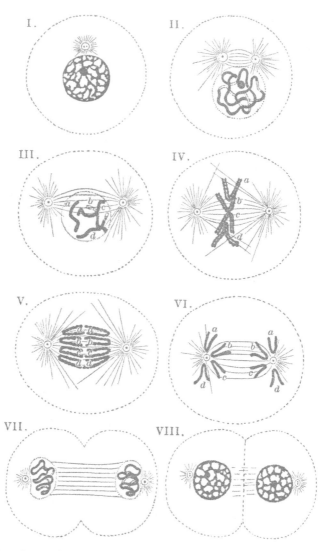

Fig. 38 a. Diagram showing normal somatic cell-division. For clearness only four chromosomes are shown. Each splits longitudinally, one half passing into each daughter-nucleus.

Prophase. I, network; II, thread.

Metaphase. III, chromosomes forming; IV, chromosomes splitting: equatorial plate.

Anaphase. V and VI, separation of half-chromosomes.

Telophase. VII and VIII, cell-division and reconstitution of daughter-nuclei.

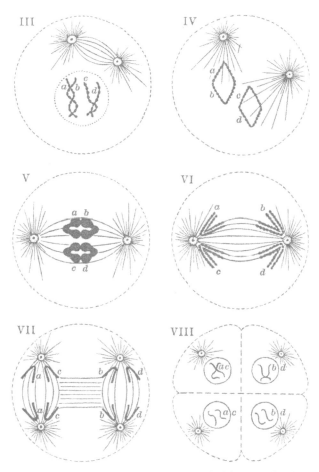

Fig. 38 *b*. Diagram showing reduction-division (meiosis) in the formation of gametes. Each gamete nucleus has only half the somatic number of chromosomes, and they are consequently not all alike.

Prophase. Much as in normal somatic cell-division.

Metaphase. III, pairing of chromosomes. IV, ring formation.

Anaphase. V and VI, separation and splitting of *whole* chromosomes.
 VII, second division into half-chromosomes.

Telophase. VIII, formation of four spermatid nuclei.

result is four gametes (three degenerate in the female), each with half the normal number of chromosomes.[1]

This description will require modification, but it will serve for the moment. What concerns us now is that we have here the perfect mechanism for Mendelian segregation. For observe: the Mendelian scheme was originally based upon the idea that the factors exist as allelomorphic pairs, such as "tallness" and "shortness" in peas. In its more modern form the presence or absence of the factor, e.g. "tallness", takes the place of the allelomorphic pair, and we regard the heterozygous form as having a single dose and the homozygous form a double dose of the factor: often we speak of these conditions as haploid and diploid respectively.

But both the old and the modern expression of the Mendelian theory seem to postulate a mechanism which must show two definite features. It must carry the factor into half the germ-cells and not into the other half in the heterozygous individual, and in the homozygous individual it must carry the factor into all the germ-cells.

This seems to demand that the vehicles which bear the Mendelian factors shall be in pairs, and that each of a pair shall be capable of bearing only one "dose" of a given factor. Thus in the heterozygote only one of the pair must carry the factor; in the homozygote both.

This mechanism is exactly what cytological investigation shows to exist. Except for certain cases to be discussed later, the chromosomes are always present in even numbers. Frequently they show differences in size and shape; and in such cases it becomes easy to see that they occur in pairs. Thus in a case where there are six

[1] For a somewhat fuller account see Glossary under "Nuclear Division".

chromosomes we might find one large pair, one medium, and one small.

Such discoveries make it certain that the chromosomes are the bearers of Mendelian factors. The meiotic (or reduction) division ensures that when the germ-cells are formed, instead of each chromosome splitting, as in the normal somatic division, and thus conveying each factor to each daughter-nucleus, the pairs of chromosomes separate, one of each pair going to each daughter-nucleus.

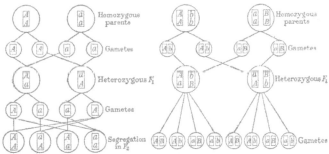

Fig. 39. These will clearly give the 9 : 3 : 3 : 1 ratio in F_2 of the second diagram if there is dominance.

Thus in a heterozygous individual bearing the Mendelian factor we happen to be considering, half the germ-cells will carry it, half will not. The observed mechanism is exactly such as will satisfy the demands of theory.

Two diagrams will make this clear. In the first we will take a cross between two individuals differing only in being homozygous for the presence or absence of one factor, in the second between two individuals differing in being homozygous for the presence or absence of two factors. We will further assume that each factor is carried by one chromosome, the other chromosomes being alike in the individuals, so that we can neglect them. It is at once obvious that the mechanism for chromosome segregation and Mendelian segregation is the same. The homozygous type has only one form of gamete; the heterozygous two forms, for each allelomorphic factor.

But we know that there are far more pairs of allelomorphic characters than there are chromosomes. In *Drosophila* (a fruit fly), which has four pairs of chromosomes, some 400 characters have been already worked out. If our explanation is true, these should show the phenomenon of linkage into four groups. Experiment fully bears this out. The factors do form four groups, and we can even tell which chromosome is carrying them.

The next question which arises, then, is whether these linkage groups are absolutely, or only relatively, permanent. Do the characters ever get transported to another chromosome?

That they do so transfer on occasion is an experimental fact. The mechanism of this transfer is known; and it even affords a rough guide to the actual position of the factor in the chromosome. Again the classical case comes from the work of Morgan and his school on *Drosophila*.

If a female fly with vestigial wings and black colour is crossed with an ordinary male having long wings and a grey colour, the F_1 generation has the expected long wings and grey body, these being dominant. Expectation gives only two, not four kinds of gamete, for both characters are carried in the same chromosome—that is to say, they are linked; and the F_2 generation should give three grey bodies and long wings to one black body and vestigial wings. Actually, though the bulk followed the expectation, there were a certain number of flies with either grey bodies and vestigial wings, or black bodies and long wings. This must mean that the chromosomes interchanged factors at the reduction division.

Of the mechanism for this process we have good cytological evidence. In many organisms when the homologous pairs of chromosomes approach and unite with one another—a stage which so far we have spoken of as ring-formation or tetrad-formation—very often another phenomenon is observed. The elongated chromosomes twine round each other, and apparently at times they break across, in such a way that when they again separate a bit of one chromosome becomes attached to the other, and *vice versa* (cf. fig. 38 b). A diagram (from East and Jones) will make this clear:

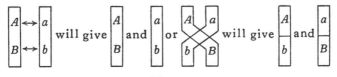

Fig. 40.

Further experiments show that the crossing-over is in this instance confined to the female. This is proved by back-crossing the male F_1 individuals with the black vestigial type, when half the offspring are grey long, and half black vestigial—the ratio to be expected if a homozygous recessive is mated with a heterozygous dominant; while if the female F_1 individuals are crossed with a black vestigial male the equal numbers of black vestigial and grey long compose only 83 per cent. of the offspring, the remaining 17 per cent. being composed of cross-over types, black long, and grey vestigial, in equal numbers.

With other characters the ratios of cross-over to normal types are different. Now it is obvious that there is far more chance of a bit near the end of a chromosome getting broken off than a bit near the middle; and so from these ratios it is possible to guess with some exactitude the position of the character in the chromosome—a truly striking result! And it is certain that the approach of the chromosomes in gametogenesis plays a part the importance of which is as yet little understood.

We may also note the possibility which this affords for the emergence of new mutations, perhaps only very occasionally if the characters are near the middle of the chromosomes, which might easily show a new combination of valuable dominants. We cannot follow up this suggestion, but it seems to me to afford a good answer to those who object that mutations cannot afford a basis for evolution, since they are usually undesirable. Granted that the fact is a fact, the *occasional* emergence of a sound mutation is all that we need as a basis of evolution.

We may now start another line of thought from the observation that a homozygous recessive crossed with a heterozygous dominant gives equal numbers of the two types.

In man, and in many animals, there are approximately equal numbers of males and females. Further, the character of sex itself is generally as clear-cut as any Mendelian factor could be. It was quickly recognised that the possibility of sex being a Mendelian character must be explored. The difficulties of such a view, and they do exist, could be postponed for the time.

The conception received startling confirmation from the work of Doncaster on the currant moth (*Abraxas grossulariata*). A rare wild variety with paler markings (*lacticolor*) was known, but only in the female. No *lacticolor* males had ever been discovered or bred. Doncaster crossed the *laticolor* female with the *grossulariata* male, and got an F_1 generation all *grossulariata*, thus proving that *lacticolor* was a recessive. The F_1 gave in F_2 the expected ratio 3 *grossulariata* : 1 *lacticolor*, but all the *lacticolor* were female. When however these *lacticolor* females were crossed with an F_1 *grossulariata* male, equal numbers of all four possibilities emerged. The *lacticolor* male had been obtained at last! But when this *lacticolor* male was crossed with the *grossulariata* female of the F_1 generation, of the offspring all the males were *grossulariata*, all the females *lacticolor*. A chart will make these facts clear:

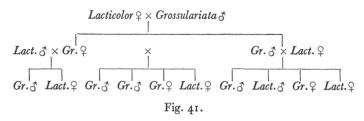

Fig. 41.

The full explanation of this complicated story need not delay us, though an outline is necessary. The really important fact is that here we have sex apparently linked in some way with other characters which behave according to the Mendelian plan. At once the suspicion is roused that sex itself must be a Mendelian character.

Doncaster's explanation of the case of *Abraxas* rested on three assumptions, all strongly supported by the evidence: (1) that *grossulariata* was dominant to *lacticolor*, (2) that the female is dominant for a heterozygous factor, absent in the male, (3) that there is a repulsion between these two factors, only effective when the *grossulariata* factor is haploid—i.e. when the zygote is heterozygous for both factors. Such an individual can then only produce the two kinds of gamete (say) **Fg**, and **fG**, of course in equal numbers. (Note that such an individual must be female, being **FfGg**.)[1]

The explanation has proved satisfactory in many other cases since discovered, though the peculiar fact emerged that in some groups of animals it was the male, in others the female that was heterozygous while the other sex carried a *double* dose.

But direct cytological evidence was quickly forthcoming. E. B. Wilson found in the bug (hemipteran) *Protenor* an odd number of chromosomes in the somatic cells of the male. The female had an extra chromosome, making up six chromosomes (three pairs), while the male had only five, one being unpaired. Further, the chromosomes were of different sizes, so that the fate of

---

[1] The symbolic table given in the second edition of Punnett's *Mendelism* might perhaps have been reproduced for the sake of completeness. Since, however, that table is based on the assumption (indicated above) that the male is homozygous for the *absence* of a sex-factor, and the female heterozygous for its presence, we omit it. It is true that in Lepidoptera and birds it is the female which is heterozygous for a sex-factor and the male which is homozygous, while in other groups the conditions are reversed, but it is very probable that the facts would be better represented by symbols which indicated that (here) the male had a double dose of the sex-character, the female a single dose. Other evidence appears to point in this direction. Our immediate purpose is served by simply noting that the conception of a Mendelian factor as determining sex does meet the case.

each pair could be followed. (All such work is done by cutting thousands of sections with a microtome, staining by special methods, and working through them in detail; or sometimes by making smear-preparations of the gonads and staining them in the same way.) It was found that while all the ova were alike, there were of course two kinds of spermatozoa, one with and one without the unpaired chromosome. Those possessing the addition, which we may call the X-chromosome, gave rise to females when they fertilised an ovum; for the X-chromosome found its pair. The spermatozoa without an X-chromosome naturally gave rise to males, since the pair was lacking.

But it is not a question of a mere bug dictating to the whole creation. Subsequent investigation has disclosed the same phenomenon of an X-chromosome in many phyla. We ourselves are not exempt. True, the cases vary a little. Some forms have two X-chromosomes in the male as well as in the female, one of the pair (called the Y-chromosome), however, being small and obviously different from the other. Either it is of no importance at all, and is merely dying away (which is unlikely in most cases), or else it carries some characters but not that determining sex. The latter interpretation is perhaps suggested by the fact that there are organisms where the X-chromosome, though recognisable, is paired in the normal way in both sexes. It is reasonable to suppose, taking the other evidence into account, that although apparently alike, one of the pair in the male carries the sex-factor (X-chromosome), and the other (Y-chromosome) does not, while in the female both are X-chromosomes and carry it—for we have modified Doncaster's hypothesis into the conception of a double dose producing one sex, a single dose producing the other.[1]

In plants the situation is a little different, for the reduction-division does not take place immediately before the gamete-formation, but in some cases (as in ferns and mosses where there is alternation of generations) much earlier. Nevertheless evidence is coming to hand of an essentially similar determination of sex in these also. An X-chromosome has been found in some cases.

Now *Drosophila* (a fruit fly) has thrown a great deal of light on this matter. If sex is determined by a definite chromosome which also carries other characters, there ought to be a group of characters which show a sex-limited inheritance. Actually, it is found that the characters of *Drosophila* fall into four groups, of which one group, which we may call A, is transmitted by the male to his female offspring only. The male *Drosophila* has only one X-chromosome, the female two, and the sex-linked characters can thus only be transmitted by the 50 per cent. of the sperms which carry the X-chromosome and give rise to females. Such cases of sex-limited transmission by one or the other sex are frequent, and study is revealing them more and more in other phyla.

One of the most widely known examples is colour-blindness (red-green) in man. It is matter of common observation that there are far more colour-blind men than women. In this case the X-chromosome, which we

---

[1] We have already noticed that in other cases the order seems to be reversed, and the spermatozoa are dimorphic instead of the ova: the results are the same; though an explanation of this strange difference is still lacking.

may now call the sex-chromosome, again carries the factor which concerns us. We may assume, since colour-blindness is recessive, that it will be masked by the dominant, so that a single dose of colour-blindness in the male will imply the complete absence of the dominant and will suffice to ensure the appearance of the recessive, since the male has only one sex-chromosome; but in the female a double dose will be required if the recessive is to emerge, because there are two sex-chromosomes.

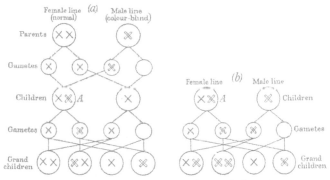

Fig. 42. (a) Normal female and colour-blind male. ✕ = sex-chromosomes; ✖ = sex-chromosome lacking red-green colour perception. (b) Female children A heterozygous for colour-blindness and colour-blind male.

Consequently, if a colour-blind man marries a normal woman, all the children, of both sexes, will appear normal; though the daughters will be carrying the recessive factor. But though the offspring of the sons will all be free if they marry normal women, half the sons of the daughters will show the defect, even though they marry normal men. If they marry colour-blind men, half the sons and half the daughters will show the defect. A diagram, partly from East and Jones, will illustrate the apparently complex, but really very simple, inheritance of a sex-linked character such as this.

This kind of evidence makes it almost certain that sex is a Mendelian factor, and that the older theories, such as that of Schenk, which attributed sex-determination to food-metabolism in the mother, and advocated a sugar increase where sons were desired, or the popular view that shortage of food in war time produces male babies, are all nonsense.

If sex is a Mendelian phenomenon, then it must be determined at fertilisation. For this there is some direct evidence. Ordinary twins are due to the presence of two or more ova in the uterus which are simultaneously fertilised; each has its own placenta and embryonic membrane (amnion). But occasionally the two halves of a fertilised egg are separated after the first cleavage, and this results in Philippine or identical twins. These share one amnion, and we have already commented on their extraordinarily close resemblance, which extends to mental as well as physical features, and persists even if the children are separated at birth, and brought up in different hemispheres. What concerns us at the moment is that such twins are invariably of the same sex. The same is true when such twins are intentionally produced by interfering with the development of a frog's egg. Again, the nine-banded armadillo (*Tatusia novemcincta*) normally produces four young within a single amnionic membrane from one fertilised ovum, and these are invariably of the same sex (fig. 43).

Unfortunately we cannot simply leave the matter here. There are some puzzling phenomena which do not easily fit in with this simple explanation. We shall shortly draw attention to the fact that Miss King was able, by selection, to alter the sex-ratio in rats by as much as 20 per cent. on either side of the normal. More

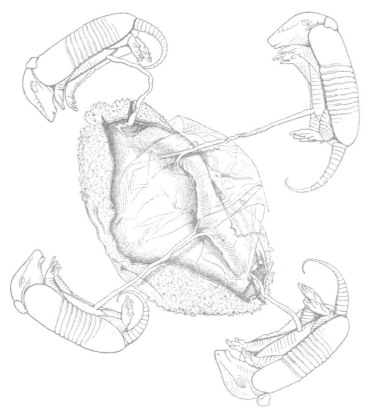

Fig. 43. *Tatusia novemcincta*. Identical quadruplets.

recently Miss King was able to produce 80 per cent. of female toads by lowering the water-content of the eggs, while Hertwig got 100 per cent. of male frogs by delaying fertilisation until the water had swelled up the yolk.

But there is another class of evidence that is still more unsettling. In 1923 was reported a case, fully authenticated, where a hen laid eggs from which she hatched the chicks, and later, owing to an ovarian tumour, developed complete male organs and functioned as a cock, fertilising eggs from which chicks were reared. The assumption of the drake's plumage (a secondary sexual character) by an old duck is well known, and all sorts of similar cases are beginning to accumulate. But as far as I am aware there is no case of a functional male assuming female characteristics, and it is perhaps a safe prophecy that no such cases will be found except in groups where it can be proved that the male has the two sex-chromosomes, and the female only one; here one would expect the possibility of the male taking on female characteristics, but not the reverse. It is easy to conceive of a hormone change which may lead to the alteration or paralysis of one sex-chromosome; while the alteration or paralysis of both might lead to death. We should therefore expect rare cases where the hormone activity leads to the destruction of one sex-chromosome, but this would only be possible in the female; and the female, having only one sex-chromosome working, might conceivably take on the characteristics of the male, in which there is normally only one such chromosome. But the opposite effect is impossible, except in the second type of organism.

Sex-reversal, then, is not in the least destructive of our general idea that sex is determined by the chromosomes.

Further, the same explanation will serve to account

for the curious alterations in sex-linked inheritance that has been produced in certain moths.

As far as secondary sexual characters are concerned, then, even the frequent adoption of the male type by females in which the normal hormones produced in the ovaries cease to function, is probably patient of this kind of explanation. And hormone change might easily affect the germ-cells in a very specific fashion, destroying certain genes, while not preventing development after fertilisation.

But the change of the sex-ratio offers a more difficult problem. Mendelian expectation would seem to indicate equal numbers of each sex; yet in few animals, probably, are the sexes actually equal. Even allowing for the fact that in many cases the male is more delicate than the female in the early stages, and dies off, there is little reason to suppose that exactly equal numbers are produced in a population to begin with. And there is this incontrovertible evidence of change of the ratio.

Two explanations seem possible. Either there might be a metabolic (possibly hormone) change which leads to the disintegration of a considerable number of the feebler eggs of one sex in the early stages of cleavage; or there might be a tendency under certain hormone conditions for the spermatozoa of one type (in cases where the two types occur in the male cells) to be more successful in fertilisation than those of the other type— a special form of prepotency. If this last were so we should expect to find the changed sex-ratio in the off-spring accompanied by a change in the ratio of other Mendelian characters. I am not aware of any evidence at present which bears upon this point. Between these alternatives we cannot yet decide, for lack of evidence;

and there are other possibilities besides. It is clear, however, that neither alternative places us under the necessity of abandoning our view that sex is determined essentially at fertilisation, and that the sex-chromosome is the essential operative cause. Indeed this is established beyond doubt. The only surprising feature is that it seems as if we need not abandon all hope of being able eventually to control the sex of offspring—if this is a really desirable thing! Probably it is, though it would mean considerable social upheaval for a time. To fear knowledge and responsibility is the worst cowardice.

Nevertheless it is well to remind ourselves that sex is not a simple thing, nor a matter of a single quality. Many males show one or another female quality in a greater or less degree, and the females also show more or less of male qualities. We have seen that in most cases a female is a male with something added; but it is not necessarily true that all the qualities that distinguish the complete male from the complete female are lodged in one chromosome, or in the sex-chromosome group at all. They may be anywhere. The only thing that does seem certain is that the most important and final distinction is lodged in the sex-chromosome: that distinction which primarily determines whether an ovary or a testis shall develop.

Much light has recently been thrown on the problem by the researches of Bridges, Goldschmidt, and Lillie, which illuminate some of the darkest corners of the problem of sex-determination. In our brief survey of these we will call the sex-chromosome X, the ordinary chromosome A, and as we have done up to now we will omit the smaller and different sex-chromosome Y, which is sometimes present as the pair to the X-chromosome. It merely complicates, without really affecting the problem.

Bridges obtained a strain of *Drosophila* in which the ratio of the A-chromosomes to the X was abnormal, and from it he was able to breed strains with varying ratios. The normal female *Drosophila* has the formula $2X + 2$ (3) A. The normal male is $X + 2$ (3) A.

Bridges was able to produce individuals having the formulae $3X + 2$ (3) A, $2X + 2$ (3) A (or with greater numbers of chromosomes in the same ratio),

$2X + 3$ (3) A, $X + 2$ (3) A and $X + 3$ (3) A.

Of these the first showed great accentuation of the female characters—super-females; the next were normal females, whatever the exact number of the chromosomes; the third formed an inter-sex, showing partly male, partly female characteristics; the fourth were normal males; and the fifth class super-males, though sterile.

This makes it perfectly clear that though the X-chromosome is the determinant of sex in the normal way, yet sex also depends on the ratio of the numbers of A and X chromosomes. Why should this be so?

It has long been known that when a cow produces twins of opposite sexes the male calf is normal, but the female frequently is sterile and in some ways betwixt and between—what is called a free-martin.

Lillie has now demonstrated the cause of this extraordinary phenomenon. The female starts all right, but the two placentas by which the embryos are attached to the maternal uterus generally fuse to some extent, thus putting the embryos into vascular connection with one another. Now the hormone secreted by the testis, which determines the general sexual development of the male, is formed earlier than the ovarian hormone, and gets to work on the heifer calf, producing some male features, before the ovarian hormone has been generated, and it also acts as an anti-body to this.

It further appears that the production of the appropriate hormone in normal cases is a function of the chromosomes, or at least is determined by factors carried by the chromosomes; and Goldschmidt has shown that different races of gypsy moths have a different potency in the production of hormones, so that by discreet blending one can produce a complete series of inter-sexes, without disturbing the chromosome-ratios at all. These discoveries in some degree solve the vexed question of the respective parts played by hormones and chromosomes, and make it possible to understand the hen who finally turns into a cock.

It is perhaps desirable to refer briefly to one case which at first sight seems to suggest that sex may sometimes be determined by nutrition. In the plant *Equisetum* the asexual spores are said to be identical. As in the ferns, these spores do not give rise to new plants directly, but germinate and form small plants called prothalli which bear the sexual organs. It is stated that in *Equisetum* the prothallus which arises from such a spore bears male or female organs according to the nutritional conditions: if well nourished the prothallus is female, if ill-nourished, male. In the ferns also there are occasional indications of a somewhat similar kind. This is undoubtedly puzzling; and the difficulty is enhanced when we remember that the reduction-division, or meiosis, occurs in the formation of the spore and not in that of the germ-cell; so that the whole prothallus generation is haploid, and if there is a pair of chromosomes bearing the X character in double or single dose, sexual differentiation ought to occur before the prothallus is formed.

Nevertheless, when we reflect that in the ferns the prothallus is normally bisexual, although the reduction-division has taken place so that the prothallus-generation is haploid, it becomes clear that we do not know the full facts about the business; and it is unsafe to build on so insecure a foundation any argument against the theory of sex-determination which we have outlined, supported as it is by so much converging evidence.

It is clear that further study is needed; and meantime it is inadvisable even to set down the possible hypotheses which suggest themselves as explanations in this peculiar case.

We now turn to a general problem of extreme interest and importance; that of inbreeding and outbreeding. It is a commonplace that a stock which has been too closely inbred grows weak. It is also a well-known fact that Jersey and Guernsey cattle, originally descended from a Norman stock, have been closely inbred for the last 170 years; no cattle being allowed to land in those tiny islands except for immediate slaughter. Almost as close inbreeding has been practised with sheep and dogs and the best horses. The evidence looks a little contradictory.

The other side of the question must not be neglected. Vigour is well known to be restored to degenerating inbred stocks by the introduction of fresh blood. The vigour and hardiness of many hybrids, such as the mule, is proverbial: in plants especially this is evidenced by large size, and vigorous habit, as in the case of our radish-cabbage cross. Yet such hybrids are more or less sterile. Hybrid vigour is a fact (the modern term is heterosis vigour); but distant crosses are usually of little practical value on account of the sterility they introduce. This is probably due to an incompatibility between the chromosomes which manifests itself in gamete-formation. We may therefore confine our attention to the crossing of different strains of the same species.

Much work has been done, especially in America, in clearing up the problem. Very little of the evidence can be given here: a good deal of it is most admirably set out in East and Jones's *Inbreeding and Outbreeding*.

Miss King inbred the Norway rat. For the first seven generations there was marked degeneration in size, vigour and fertility. Some of this was due to unsatisfactory feeding, for the controls suffered in the same way. After that, rigorous selection was made, only about 2 per cent. being allowed to breed. There was a definite advance in vigour and fertility! A further point of interest was that, by selecting those which tended to produce an excess of one sex in the offspring, the normal sex-ratio, 105 males to 100 females, was altered by about 20 per cent. in either direction.

Rommel's experiments on guinea-pigs gave rather a different result. There was at first a marked deterioration; afterwards some of the selected lines went on well, others degenerated still further, becoming less fertile, and less vigorous.

The general result of a great deal of work tends to show that what happens has nothing to do with inbreeding *per se*, but simply depends on the characters already in the stock.

Essentially, inbreeding brings about the segregation of Mendelian factors. Thus it will clearly tend to bring out recessive characters; since there will be a larger percentage of forms homozygous for this or that quality. If the quality concerned is a bad one there will be deterioration: if it is a good one there may be improvement (*may* be, because the vigour of a stock depends, not upon one or two, but upon many qualities).

We have said that recessives will tend to emerge. Is

there any reason for supposing that recessives are more likely to be harmful than dominants? Certainly there is. A dominant is more or less effective in the haploid state; a recessive is masked, and may not affect the carrier at all in the haploid state, though probably in many cases it is slightly disadvantageous.

A disadvantageous dominant will be quickly weeded out by natural selection; but since the recessive will only appear in the diploid state this will not be so weeded out. Many individuals will carry it as a latent character.

If, then, many or most mutations are due to a fresh segregation of Mendelian factors it stands to reason that the majority that emerge are likely to be recessive characters; and since the recessives have not been weeded out, many of these, probably most, are likely to be bad. The frequent weakness of mutations is thus easily explained.

But there is no reason to suppose either that all mutations are due to the emergence of recessives, or that all recessives are bad; though probably most are. Emergence of a new dominant combination might arise, for instance, by the transfer of a group of characters, say in the male, to a different chromosome, by a cross-over, which brings them into close association with a different set of characters in the female chromosome corresponding.

Those Lamarckians who base their objection on the weakness of most mutations are on shaky ground. What they have to prove is that there is never a strong mutation.

We will defer the consideration of heredity in man till later, but I cannot refrain from drawing attention here to the interesting example of the Egyptian dynasties which illustrates the phenomenon of Mendelian segregation brought about by inbreeding.

Owing to the matriarchal inheritance of the primitive Nile tribes, and the subsequent domination of tribes with patriarchal inheritance (such is the probable explanation), it became customary for the dynastic pharaohs to marry their own sisters or half-sisters, thus ensuring the legitimacy of inheritance against all objection. In the Egyptian dynasties we see the result which we should expect from such inbreeding. Desirable qualities appear, and the dynasty rises to magnificent heights; then follows the emergence of undesirable recessives, and the dynasty falls a prey to internal revolution or foreign invasion, and collapses with dramatic suddenness. A famous example is Aknaten or Iknaton (Amen-hetep IV). His mother, Tyi, was a Syrian, but of the Egyptian Blood-Royal; for towards the end of the Eighteenth Dynasty intermarriage with Syria was customary. The son is a religious and artistic genius of the most outstanding kind, but there are physical defects, if we may judge by the caricature-statues in the Cairo Museum; and the Tell-el-Amarna tablets point to an almost incredible political ineptitude, as indeed does his quarrel with the priests, so weakly conducted that it led to a forced removal of his capital. Generations of inbreeding have done their work. Qualities desirable and undesirable have segregated and emerged.

We may now turn to the other side of the question—outbreeding and heterosis vigour. Here we enter upon ground where the very simple conceptions with which we have been dealing must be supplemented. For Mendelian experiment comparatively trivial characters, which are easily recognised but which do not vitally affect the health of the organism, are chosen. Even of such easily recognisable characters over 400 have been traced in *Drosophila*, which has only four chromosomes. It is therefore clear that a chromosome carries many characters: as a first guess we may suspect one, or a closely allied group, to each chromatic granule or chromomere. Of these, some will be the superficial characters with which we have largely been concerned; others will be determinants of the metabolism of a tissue, or of the whole body.

Combinations of such genes may in some cases be very bad or even lethal. A well-known example of this arose from the attempt to breed a homozygous yellow mouse. It was found that the ratio worked out as 2 yellow : 1 normal (yellow being a dominant), instead of the expected 3 : 1. But in the uterus of the mother were found the missing embryos in a disintegrated state. A double dose of yellow had upset the metabolism of the organism completely, and it could not survive.

It is more than probable, then, that there are many genes which we shall never be able to recognise: the organism will be healthy or unhealthy, and that is all we shall know. In such cases it is very likely that there will be imperfect dominance; and imperfect dominance means, in many cases at any rate, that the more or less recessive character is not without influence in the hetero-zygote. Indeed it may be suspected that it is not without influence even when there is dominance, for dominance is in fact rarely perfect, and it may be questioned whether a single dose of a dominant factor can ever be as fully effective as a double dose.

On the other hand, this by no means necessarily implies that a single dose of a factor will not be pretty effective if it is a dominant, or partial dominant. We know that this is in fact true.

Bearing these things in mind, we may consider an imaginary case of crossing two individuals each containing three pairs of chromosomes, of which each pair bears three homozygous characters, all of which are dominant. Further, we will assume that these two individuals carry entirely different sets of characters, three to each pair of chromosomes, and that each character represented by an odd numeral can be replaced by a

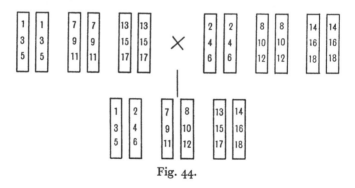

Fig. 44.

character represented by the even numeral next above it, and *vice versa*. Since there is full dominance, the characters will be as effective in the haploid as in the diploid state. The diagram (modified from Jones and East) illustrates the result.

It is clear that the hybrid F_1 contains twice as many genes as either parent. As all are dominant, all are effective. It is reasonable to suppose that, since the nature of an organism depends on the sum of all its genes, an organism with all these dominants will be more vigorous and developed than either parent form.

(We must of course assume that each parent has in addition a stock of genes, common in nature, which will afford the groundwork for development.) We should, in fact, expect heterosis vigour in F_1. It is unlikely that an exactly similar constitution will occur again; the chances are enormously against it, for the F_2 generation brings segregation; but most of the offspring will carry some additional characters, and thus have an advantage over the original race.

Though this illustration presupposes an extreme case, it enables us to catch a glimpse of the meaning of heterosis vigour. In many instances heterozygosis may bring definite advantage. If the form is homozygous for certain fundamentals, and has a number of additional advantageous characters in a haploid state; and if, further, such forms are interbred; there will be no danger of losing the essentials, while the changes will be rung on a number of other desirable qualities. Vigour will be maintained, and there will be a good deal of variation, sometimes highly favourable.

Such an idea is fully borne out by experimental work; and we may look on the problem of inbreeding and outbreeding as solved in principle at least. It is easily summed up.

Inbreeding is neither bad nor good: its effect depends entirely on the genes carried by the race. Inbreeding merely produces segregation of both dominants and recessives. Outbreeding does the reverse—it produces heterozygosis. Again, whether this heterozygosis is good or bad depends ultimately on the stock of genes, though ill effects are masked. Recessives do not appear; but it is rash to assume that they are entirely without effect.

Obviously, from the breeder's point of view the right

thing to do is to inbreed in order to eliminate the weak genes, then to cross the good inbred stocks in order to acquire heterosis vigour, and then to resume selective inbreeding.

This can only be done where rigorous selection is possible. In human beings, since it is socially inadvisable to throw out the weak offspring, it is a plain duty to prevent their occurrence. Therefore first-cousin marriages are to be deprecated where undesirable characters appear on either side of the family tree: they are only justifiable between really sound stocks. Similarly, any marriage into an unsound stock is highly risky, for it is pretty certain that some of the descendants will reveal the terrible recessive. Of this we shall have more to say later.

The Knight-Darwin Law of Intercrossing has now assumed a different complexion. Inbreeding is not in itself a disadvantage; but what the anthropologists call exogamy—marrying outside the clan—is generally advantageous, for reasons that are no longer obscure, but mathematically exact.

Finally we may turn aside from our main subject for a brief moment to consider how conjugation, which is the starting-point of exogamy, can have originated. This is clearly a separate problem: the mechanism for the *determination* of sex does not explain the *origin* of sex. Once there, the advantages of sex are patent. Besides potentially adding to the vigour of the organism it affords a vast range of possibilities of variation. Suppose there are twenty variations in an asexual species: that gives just twenty variations and no more. In a sexual organism 2^{20} or 1048576 varieties are possible, though very improbable. Evidently there is overwhelming advantage in the sexual process, once it is established. But this throws no light at all on the actual establishment of the process.

An observation which does seem suggestive on this point is that the earliest manifestations of conjugation, at all events in plants, and probably in some protozoa and even in higher types, seem to be associated with

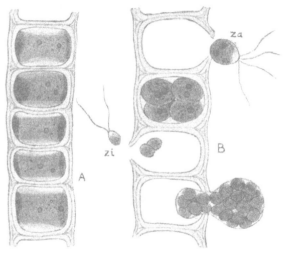

Fig. 45. *Ulothrix zonata*. A, algal filament; B, filament showing 4-ciliate zoospores *za* and 2-ciliate swarmspores *zi*.

unfavourable conditions, and especially with the failure of the food-supply. In the lower animals it is also associated with a tendency for the waste products to accumulate and set up some form of poisoning, as in *Paramoecium*. When food is plentiful, a simple alga like *Ulothrix* (fig. 45) forms spores, which swim away and spread the species; but in times of scarcity smaller spores are formed, and when two of these meet their cilia become entangled and the spores fuse. This may be looked upon as a form of cannibalism—a chemotropic reaction brought about (probably) by nitrogen-hunger.

Nevertheless such a phenomenon gives at least a probable origin for conjugation. We know little of the physics of the nucleus, but it is by no means inconceivable that when two similar organisms fuse in this way some force, physical or chemical, might draw the similar chromosomes together; for we see such a force at work in every fertilisation, and in every reduction-division.

Is it fantastic to guess that in nuclear phenomena of this kind we are watching the perfected outcome of the very primitive phenomena which attended the first cannibalistic conjugation? Is it fantastic to guess that this cannibal meal gave such renewed strength as enabled the weakened organism to get rid of poisons which had accumulated through an incomplete metabolism due to weakness, and to find here a possible explanation of the fact that in a depressed stock of *paramoecium* vigour is slowly recovered after conjugation?

The survival-value of such a cannibal habit is obvious, and a line tending to conjugate would quickly be established. Moreover, the diploid individual, with its vast potentiality of variation, is in a strong position both for establishing the habit, and for evolving. Conjugation, as we have said, is a big asset.

Obviously severe measures will have to be taken for the throwing out of unwanted nuclear material, but here again the survival of those which do this, and the quick extinction of those which do not, is easily understood.

One odd little point is perhaps deserving of mention, though no decided opinion can be expressed—telegony.

Many breeders of horses and dogs are of opinion that if a valuable mare or bitch has a *mésalliance*, the future offspring by suitable sires will be affected. The classic instance is Lord Morton's Arab mare, which was in foal by a quagga (a close relation of the zebra). After a subsequent mating with a blood-horse, the foal was born with oblique black stripes on the hind legs. It is stated, however, that this does occur occasionally in the foals of Arabs and Barbs without any such history; the markings fading out as the first coat is shed. I have recently come across a similar case, by hearsay, in Devonshire. The evidence, though not exact by scientific standards, is quite good. In this case the mare (Arab) had had several foals with no markings, then a foal by a zebra. The next foal by a horse showed zebra markings. Even if the case is correctly reported, the evidence is not conclusive, for it is quite possible that only old mares have foals of this (possibly) atavistic type.

The matter should not, however, be dismissed as an impossibility, though one is not inclined to attach much credence to it at present. It must never be forgotten that there is a good deal of cytoplasm, as well as the nucleus, in the ovum, and a little in the spermatozoon; and it is not even unlikely that a cytoplasm modified by hormones may exercise an influence on the developing organism, and so constitute a secondary source of what may be loosely called hereditary characters, by handing on a transmitted environment for the nucleus. We have seen evidence of this in the work of Bridges on *Drosophila*, and of Lillie on the free-martin. It is even conceivable that this influence might last long enough to affect the genes, and so to give a foundation for a form of the transmission of acquired characters. This, in the present

stage of our knowledge, is no more than a wild speculation, though the recent work just mentioned is in its
favour: probably the wise attitude is that of agnosticism,
tinged with scepticism.

After all, there is some evidence for the effects of
prenatal influences initiated in various ways, though it
is not conclusive and probably holds only for a very
generalised effect. Such influences we are inclined to
regard as environmental: the idea of a changing environment for the nucleus may not prove unfertile, and
seems to offer a bridge whereon the rival Darwinian and
Lamarckian kings may meet and make a treaty.

Chapter VI

INHERITANCE IN MAN

Mendelian heredity has been proved to hold good in the case of man over and over again. Analysis is only practicable when not more than one or two factors are concerned, since we cannot have recourse to breeding experiments; but family trees supply ample evidence in a good many cases. Only a few will be mentioned here.

Defects and abnormalities supply excellent illustrations. Some are dominant, others recessive; a few show complications as yet unexplained. Some are sex-limited, like colour-blindness, others are not. For instance, brachydactyly, a condition in which the fingers have only two joints instead of three, is a simple dominant. (St Mark may have had this abnormality, though κολο-βοδάκτυλος may merely mean that he had lost a finger.) Haemophilia, in which the person concerned may bleed to death from a slight wound, owing to the absence of the ferment which causes clotting of the blood, is a recessive, but nearly sex-limited to the male, for obvious reasons. Diabetes insipidus, alkaptonuria, haematuria, presenile cataract, ophthalmoplegia, nystagmus, absence or defect of enamel in the teeth, tylosis, oedema, night-blindness, are a few among many cases of Mendelian inheritance of defect in man. The good teeth of the Italian and the bad teeth of the Irish are said to be as conspicuous in New York as in their native countries. (Here the difference is a question of the enamel.)

Of the inheritance of normal characters there is far less evidence; the fact being that such characters are not so markedly delimited as a rule, and usually involve a number of factors. It is, however, clearly established that brown eyes are dominant to true blue. A true blue eye has no brown pigment at all, but greys have brown pigment only on the retinal side of the iris. In fact, all eyes, except those of albinos, are blue; but brown pigment may be added; and the heterozygous brown is markedly different from the homozygous, giving greys and greens. The expected ratios emerge when a number of cases are tabulated, though complications do exist.

Again, elliptical section often is dominant to circular in the hair. The former gives wavy hair, the latter straight. An additional factor flattens the section still more in the "wool" of the negro.[1] Hair-colour is more complex, though flaxen seems recessive.

Skin-colour is interesting, though not fully worked out. The negro appears to have at least three factors for blackness, and probably more. Thus, if the negro be written B_1B_1, B_2B_2, B_3B_3, and the white b_1b_1, b_2b_2, b_3b_3, then the mulatto will be B_1b_1, B_2b_2, B_3b_3. But a possible condition is B_1B_1, b_2b_2, b_3b_3, which might mate with b_1b_1, B_2B_2, b_3b_3, when the offspring would be B_1b_1, B_2b_2, b_3b_3; probably much the same colour as the parents, but not necessarily so; and the next generation would be likely to show offspring both lighter and darker than the original. The idea of a black baby arising from the marriage of a pure white with an octoroon, or even a mulatto, is impossible; but the possibility of a black baby from the marriage of two very slightly coloured parents is a fact.

As further confirmatory evidence, in a general sense, that human heredity follows the normal lines, it is worth

---

[1] Inheritance is irregular and appears more complex than is suggested here. Baldness, also, is dominant in the male, recessive in the female.

noting that Karl Pearson found the same kind of intensity of inheritance—approximately the same coefficient of correlation (see p. 80)—for both, not only in physical but also in mental factors, as we shall see; and stature and ability alike gave to Galton the curve of normal probability in human beings. (We have noticed already that these curves approach perfection more nearly the more factors there are involved. Thus we should expect good curves both for stature and ability, together with a practical impossibility in analysing them into their Mendelian factors.)

We have, then, ample evidence to justify our opinion that Mendelian laws do not express a fate laid only upon the lower animals. Man is equally subject to the conditions they formulate. And this conclusion is of great importance. No doubt complexity is greater, and other factors intrude more conspicuously: we must not overlook this; but to overlook the part played by Mendelian factors is an error far graver.

Nevertheless it is unjustifiable to represent the facts as more simple and clear-cut than they really are. The more closely inheritance is studied, at any rate in man, the more clearly does the fact emerge that the Mendelian ratios offer only a first approximation in many cases. Nor should this surprise us. Man owes his position, biologically speaking, to the fact that he is the most viable, the most adaptive of organisms; and this must imply that he not only carries a vast number of genes but probably that they are more interdependent than is the case in simpler and less viable organisms.

In such conditions one would expect that a given characteristic should often be the resultant of two or more genes which, separated, come out in a form rather different because unmodified by the presence of the other. Moreover it is possible, though by no means necessarily true, that hormone balance and other metabolic conditions will have more effect upon interdependent genes in a complex viable organism than in the simpler animals.

In brief, while we are bound, I think, to regard the Mendelian phenomena as being fundamental in man in the same degree as they are fundamental elsewhere, we are not debarred from believing that the more complex and viable the organism the more likely is it that the influence of other and more superficial factors may mask or even in some degree modify the basic phenomena.

Even if in many cases we cannot set out our schemes of inheritance in simple Mendelian form, owing to the complexity of the factors, there is nevertheless every reason to suppose that heredity plays as important a part in the physique of man as it does in other animals. Simple cases are definitely Mendelian sometimes, and it is only the difficulty of experiment that prevents our analysing the factors in detail. When out of 304 marriages of affected persons producing 1012 children 589 of these are found to be affected with presenile cataract even in infancy, there can be little doubt of the hereditary nature of the trouble!

But it must be remembered that we often call by one name a whole lot of different conditions. Alcoholism is a group of several distinct conditions: deafness is at least twenty different diseases. We must be very careful to diagnose the condition rightly before we begin to consider its inheritance.

Another point of interest arises as soon as one begins to study the heredity of disease in man. Certain diseases which are definitely bacterial in origin appear to be inherited. In one or two cases there is actual infection of the germ-cells (e.g. syphilis) and true heredity is not concerned; but in most this is not so. Evidently what we have here to study is proneness to take the disease; and on examination this generally, if not always, resolves itself into lack of immunity. We may therefore digress a little from our main theme to study briefly this question of immunity.

When we look at the case of consumption—pulmonary tuberculosis—it is difficult to resist the conclusion that it tends to run in families. Yet tuberculosis is a bacterial disease, and is acquired independently by each sufferer. Moreover, the chance of infection within the family is obvious, once it has appeared, so that we must not jump straight to the conclusion that the tendency to consumption is hereditary. On the other hand it is a disease of young adults, rather than of children; and this fact in some degree discounts the likelihood of infection in the nursery days. Also, it is not often passed from husband to wife or *vice versa*. Again, in theatres and cinemas everyone is liable to be exposed to infection. The disease is growing less prevalent,[1] and this may be attributable in part to more healthy customs, such as sleeping with open windows, and taking open-air exercise. None of these facts must be neglected; yet the evidence from family trees is strong.

Though the disease itself cannot be hereditary, the

[1] This was true until lately, and is still true of the population of England as a whole. But of late in the poorer section of the upper classes there has been an increase, attributable to the fashionable slimness, and to expenditure on motor cars and amusement which would have found a better outlet in buying healthy food.

power of resisting it may be; and this clue seems to make the contradictory evidence fall into line. Those with no resistance catch the disease and die, since all town-dwellers are exposed to infection at times. Those with some resistance who lead a healthy life are able to resist it. Thus an immune, or partly immune, population may be established, and the incidence of the disease will grow less. Lack of immunity seems to run in families—here is probably the hereditary factor. In the nineteenth century, when great migration to towns began, the disease became very prevalent; for in the country it is possible to avoid infection, and the open-air life prevents the disease from taking such a firm hold as in towns. By the twentieth century the remains of the population is becoming relatively immune. The non-immune country population has been sifted out in the towns.

What exactly do we mean by immunity? When a patient is recovering from typhoid, the recovery is due to the secretion of an anti-toxin by the body-tissues, which not merely neutralises the poison generated by the bacillus, but actually destroys the bacillus itself, or leads to its destruction after it has been rendered harmless. For the future the patient is nearly immune from further attacks by an *acquired* immunity: the bacillus can generally get no hold.[1] This formation of anti-toxins lies at the bottom of all our vaccination and serum-therapy. The result of it has been that whereas in

[1] Such immunity from future attack is not *generally* characteristic of the bacterial diseases, though it is characteristic of most virus diseases (diseases caused by organisms so minute that they pass through filters which will stop the smallest bacteria). The fact is interesting and at present unexplained; but at the moment we are discussing only the cases of hereditary immunity to any disease without reference to the actual nature of that disease.

the Boer War some 50,000 cases of typhoid occurred in the English armies, in the Great War, among ten times as many men, far more closely packed, there were, I believe, nineteen cases all told. But there are people who have a *natural* and inherited immunity to this or that disease; and in course of time a population tends to get at least a partial immunity to a prevalent disease; so that it becomes rare for an attack to be fatal. When the Crusades and the Renaissance brought syphilis to England the results were appalling; the grave-digger in Hamlet has something to say on the matter. The disease was new to the West, and there was no immunity. To-day we find Polynesian and Pacific islands almost depopulated by measles, a disease unknown there till the coming of white men. The destruction of African and North American races by syphilis and alcohol is well known.

The last affords rather an interesting illustration of another type of immunity (though ultimately we must regard all immunity as a kind of resistance to the poisonous action of some definite chemical substance which tends to upset the metabolism of a cell or tissue, without reference to the way in which that substance originates). There are at least four types of alcoholism sufficiently different to rank as separate diseases. Among these four one form of the rare disease, true dipsomania, appears to be a variant of suppressed gout, for instance, and yields to treatment with Vichy water when the rhythmic craving is foreshadowed by the characteristic depression; though other cases of true dipsomania are not helped by this treatment. But it is the disease known as false dipsomania which for our purpose is of the greatest interest. In this disease there is no craving until it is aroused by a minimal dose of alcohol—the necessary quantity may be very small indeed. Once aroused, the craving is irresistible, and does not cease until the patient is too ill to drink any more. Total abstinence is perfectly possible, and is the only hope in such cases.

The explanation appears to be—though positive confirmation is lacking—that alcohol always induces the secretion of an anti-toxin. In the pseudo-dipsomaniac a minimal dose induces an over-secretion, which produces the craving, since the body seeks to neutralise the anti-toxin. The further dose taken to satisfy the craving again causes over-compensation, and so it goes on. An extreme over-dose of anti-toxin brings on delirium tremens: hence the danger of suddenly cutting off all alcohol, instead of tapering the doses: this is however of more importance in cases of the two chronic types of alcoholism.

In the normal moderate drinker, it is probable that only the requisite amount of anti-toxin is produced.

Racially this is of great interest. It is well known that drunkenness is comparatively rare in wine-producing countries. The population is immunised to moderate doses of the poison (to large doses of concentrated alcohol no population is immune, so that alcoholism due to spirits may be found, though even that is not very prevalent).

In countries where alcohol is not plentiful there is no such immunity, and spirit-drinking especially leads to disastrous results. The population is not immune.

Immunity is probably ingrained in a race by a process of elimination. The weaker germ-cells are probably killed—of this we have already seen evidence in rats, etc. —and the individuals with a tendency to over-produce the anti-toxin drink themselves to death. If the tendency be a recessive factor, as seems likely, there is little hope of eliminating it altogether; but gradually a race which is predominantly immune will be developed. It is worth noting, however, that if selection ceases, the population will gradually lose its immunity, and in a few hundred years will be as open to the dangers of alcohol as the North American Indian. As it is impossible to eliminate alcohol from the world, one could not regard a successful attempt at prohibition with anything but alarm; racially it might in the long run prove far more disastrous than the cocktail habit.

Immunity is specific within the narrowest limits. The anti-toxin is one particular globulin, which neutralises the effect of only one particular albumose (the toxin) as a rule. A man may be immune to the common boil-bug, *Staphylococcus aureus*, yet highly susceptible to his equally unpleasant variety *citreus*.

After this digression we may return to our main business, the discussion of heredity in man. We may reiterate the statement that what is true of bodily conditions seems to be equally true of mental. Karl Pearson, working statistically, finds the coefficient of correlation between parents and offspring, and between brothers and sisters, to have the same range and average value for mental as for bodily factors. Galton got as good a normal curve from the marks obtained in the Mathematical Tripos as he got from his chest-measurements of Scottish soldiers. But in mental qualities there is likely to be much interdependence of genes, and one would hardly hope for many instances of simple Mendelian inheritance. The fact that broad statistical treatment leads to the same results in the case of mental characters as in the case of physical ones is fairly good evidence that the underlying causes are the same. For at least they indicate a similar intensity of inheritance in both. But in some cases we do actually appear to find indications of simple Mendelian ratios.

We must now turn to the inheritance of ability and of mental defect—always remembering that the latter may not be due to the same cause in all cases. A slight pressure on the brain, a gross malformation, an injury, a bacterial infection, a hormone disorganisation, will each cause mental derangement; not only the hereditary factors that we are going to consider.

The first thing that emerges is that men are not all equal mentally. The men who get honours in mathematics are all above the average, since all get an honours degree; but the senior wrangler gets on the average thirty times as many marks as the bottom man—this on papers with a time-limit such that no man can hope to do nearly all the questions, and where the bottom man has done all he can, while the senior wrangler could perhaps have done much more.

In 1869 Galton published his *Hereditary Genius*, an investigation of pedigrees, and Havelock Ellis, Cattell, and many more have since added to our knowledge. The evidence of such investigations is conclusive. There can be no manner of doubt that if you want to be reasonably sure of possessing ability you must choose able parents.

There is a good deal of evidence in favour of the inheritance of a particular type of ability; but there is also evidence of the inheritance of more generalised ability which can branch out in various directions.

Of the first, many instances could be given. Galton found in eight generations of the Bach family no less than fifty persons showing eminence in music. This eminent musical ability increased for the first four generations, culminating in Johann Sebastian Bach, and then gradually diminishing. It is worth noting that musicians tended to marry within the guild; so that the faculty was probably imported from both sides, at least in the earlier generations.

The Wesleys afford a pretty example of two qualities linked through several generations—music and religious fervour. Two of the Wesley grandfathers were expelled from their benefices as non-jurors. Charles and the great John showed both qualities in a high degree. Then comes Samuel Sebastian, a great composer of church music. His son, who died fairly recently, was a clergyman whose passion was music.

A study of the peerage, or of the *Dictionary of National Biography*, brings home the descent of a particular type of ability through many generations. Scropes and Boyles, Cannings and Moncktons, Noels and Gages, Darwins and Barings—wherever one looks one finds the same thing. Generation after generation serves the State with distinction, and usually in the same kind of way. One family will show military ability, another civic and parliamentary, another statecraft, another inventive, another scientific. In *The Family and the Nation* by W. C. D. Whetham a considerable amount of interesting material may be found; but anyone who will take the trouble can find fascinating occupation for a wet afternoon in routing out the facts for himself. Only a peerage and the *D.N.B.* are needed for a very convincing demonstration.

So too with less eminent professions. Woodcraft, shepherd-craft, smithcraft, run in families. Devon still supplies a large proportion of the men for the Royal Navy; and men of the sea follow the sea for generation after generation. An obvious objection to such an argument is that it is all a question of environment. A man brought up in a ruling family, a military family, a seafaring family, is likely, through opportunity and suggestion, to follow the family road. This is true; but study of family histories over many generations suggests that it is not only a question of environment. The sons of great proconsuls have a chance to enter the Services, for instance, and do enter them; but the men who really stand out in the family are generally the rulers rather than the fighters; and as the stock weakens we shall, I think, find plenty of meritorious but uninspired soldiers springing from it, but fewer and fewer men who rise to leadership in the old field; and it is unlikely that a Marlborough or a Napoleon will arise among the soldier descendants. The great general has a combination of qualities, all developed in a high degree: he must thus be one of the rarest examples of genius.

The broader question of whether the apparent inheritance of ability might not after all be due to the advantage of environment, was carefully investigated by Galton who proved once for all that this explanation will not hold water. For this purpose he took the judges of England, since a judge attains that eminence only by merit, and family connections are of singularly little advantage to him. In order to understand his conclusions it is necessary first to make clear what Galton means by eminence, and what proportion of eminent men is usually found in the population of England.

Assuming that all eminent men will have made their mark by fifty, he first found the proportion of men of that age who had biographies in *Men of Our Time*, the less catholic version of *Who's Who* then current. This would seem to offer a fair criterion of eminence. The proportion worked out at 0·025 per cent. of the male population of that age. He then made a similar investigation of the obituaries in *The Times* of 1869, considering only obituaries of two inches or more. The proportion was approximately the same. Allowing for the smaller population, an investigation of the obituaries of earlier years gave the same figure. More recently investigation of the *Dictionary of National Biography*, covering all ages, but representing a higher standard, disclosed a somewhat smaller figure—about 0·02 per cent. Thus it is reasonable to say that 250 in every 1,000,000 of the population are of eminent ability.

Galton then examined the family records of the judges between 1660 and 1865, the premiers of the last century, and the statesmen of the reign of George III. For the 286 judges he found the following numbers of eminent relations. Great-grandfathers 0·2 per cent., grandfathers 2·6 per cent., fathers 9·1 per cent., uncles 1·6 per cent., brothers 8·2 per cent., first-cousins 0·5 per cent., sons 12·6 per cent., nephews 1·7 per cent., grandsons 3·7 per cent., great-nephews 0·7 per cent., great-grandsons 0·5 per cent.

When we remember that the percentage for the whole population is 0·025, the figures show a phenomenon far beyond the range of accidental error, or of defect in the method of computation. Set out in tabular form, the effect of close relationship to the eminent men with whom the investigation starts jumps clearly to the eye.

His other investigations led to the same kind of conclusion.

Without spending further time on individual cases we may admit that the evidence for the inheritance of ability is overwhelming, and that for the inheritance of ability along specialised lines it is strong; though it is not possible to *prove* that a man who has done good work in one sphere of life would not have done equally good work in another. Still, investigation of pedigrees does show that one family does pre-eminent work in administration, another in local or parliamentary affairs, another in the army, another in science, and so on; and that few members of those families rise to any great height in alien fields. Scropes and Cannings, Boyles, Barings, Darwins, Rothschilds; each has its own line. And so with countless others. The Darwins, with their relations the Galtons and Wedgwoods, are particularly striking. At least sixteen men of scientific eminence, more than half of them lineal descendants of Erasmus Darwin, arose in this family in five generations, of whom nine or ten were fellows of the Royal Society.

Dilution alone is not responsible for the dying out of such desirable germ-plasm. Throughout the Middle Ages the ablest men with hardly any exception became clerics. Celibacy of the clergy must have been responsible for a disastrous extinction of talent. Until the early 'eighties Fellows of Colleges were not allowed to marry; if they did so they had to give up their fellowships. This restriction weeded out a vast amount of ability from the race: its removal is already producing a marked effect upon the class lists and scholarship rolls of the older Universities.

Another way in which we have eliminated conspicuous ability is by honouring its possessor with a peerage, which carries dignity and increased expenses, but no money. The only hope for the next peer is to marry an heiress. Now in nine cases out of ten, at any rate before the industrial expansion of America—a period now probably drawing to its close—an heiress came from a sterile stock. Unless several tributaries had flowed into the main channel it was not likely that the heiress would be really rich; and she would certainly be an only child, or one of two, in days when large families were the rule. A study of the subsequent history of those raised to the peerage convinced Galton that the extinction of the title could generally be traced to a marriage with an heiress which introduced a factor of sterility. In every case except one marriage with an heiress did actually result either in the complete extinction of the family or in the narrowest of escapes from such extinction, the situation being just saved by another marriage into a fertile stock.

It is interesting to notice that in England there have been three great outbursts of ability since the Conquest; in the thirteenth, sixteenth and nineteenth centuries. Then it was that new families came to the fore. Ten generations apart we find the Crusades and the conquest of France, the Elizabethan expansion, and the Industrial Revolution. A biologist does not look on such a series as chance: he wonders whether there is any other explanation.

There was a great mixture of nearly-allied populations at the Conquest; and it is possible that ten generations gives about the proper time for the segregation of the mixed germ-plasms. Of this we shall say more later. The new families will intermarry with the old, causing a fresh set of germinal combinations; and the addition of French and Low Countries blood (which happened to take place to a considerable degree) may easily account for the second and third crests.

To the third, however, another factor chiefly contributed: the Yorkshire dalesmen, a magnificent yeoman stock, utilised their water-power, turning from hand-weaving to machinery, and rose rapidly to wealth and intermarriage with the old outstanding families. It may be that this admixture, and others which increased facilities of transport have brought about, may provide a new crest in 2100–2200. Till then it is not likely that England will do anything very great; and the last quarter of a century has seen the initiation of changes which render even the most cautious of prophecies singularly insecure. And all these are minor admixtures, very different from the great ones which found a new race.

Of one thing alone the biologist can feel certain: it is upon the germ-plasm of the nation that its future depends. The destruction of the bourgeoisie, either by revolutionary methods, or by a selection of the unfit imposed by mistaken political developments, quite certainly means that the nation is doomed to insignificance. But this matter will be considered in due course, and we need not anticipate.

Flinders Petrie, in a little book entitled *The Revolutions of Civilisation* (a book whose significance seems to have escaped the notice of biologists, no doubt owing to the fact that it is primarily archaeological in interest), finds a striking series of rhythms in the history not only of Ancient Egypt but also of Europe and other continents. He brings forward strong evidence to show that a climax

of artistic ability is reached some 700–800 years after a
conquest or invasion by an alien race, a climax of
mechanical skill a good deal later, and of wealth a little
later still. The intervals between the successive climaxes
lengthen in later ages, and it may be unwise to lay much
stress upon them, but the order is always the same, first
art (and sculpture precedes painting, painting precedes
literature), then mechanics, then wealth. The really
suggestive fact is that these rhythms do apparently occur
with the first maximum at 700–800 years after a racial
admixture. As Professor Petrie pregnantly observes,
"Parthenogenesis is unknown in the birth of nations".
England suffered her great invasion somewhere about
A.D. 450 550. The thirteenth and fourteenth centuries
showed the artistic climax, the fifteenth to the seven-
teenth the literary, the nineteenth the mechanical. With
this phase comes wealth followed shortly by decay, if the
familiar rhythm is to be repeated.

Social, political and economic convergence, brought
about all over the world by increased facilities of transport, have so altered our conditions that it is rash to say
now that what has happened in the past *must* happen
again in the future. Yet it would be foolish to deny that,
though detailed conditions may alter, the basic laws
which govern biological rhythms remain and cannot be
inoperative.

Having learned to control so much of our environment we have introduced new factors. But in the past
a striking phenomenon of the period of decay has been
the extinction of sound germ-plasm. We shall shortly
turn to the question whether there are indications of any
similar phenomenon in our own day; and we shall have
anxiously to admit that they stand out only too clearly.

For England there is still a disastrous possibility, as we have already hinted. As prologue to its consideration we may now turn to the inheritance of undesirable characteristics.

The family histories of the children in special schools for the mentally deficient very frequently show that the trouble runs through the family. Such a case is the following. Father unknown, mother mentally deficient. The daughter is mentally deficient and marries a man whose mental history is unknown, but who obviously carries an undesirable factor. Offspring four: a daughter, prostitute before fifteen years of age; a daughter, mentally defective, not married, who has a son, mentally defective, in a special school; a daughter, mentally defective, in a special school; and a son, said to be normal.

The brothers and sisters of 150 children in special schools were tabulated. The birth-rate is high among such families, and there were 1269 of them. Of these 813 were either stillborn, died young, were mentally deficient, or were criminals or paupers. Only 456 were said by their parents (whose standard would not be high) to be normal. Boards of Guardians know only too well that children born in the workhouse are liable to be of low mental grade. The same feeble-minded woman, incapable of self-control, returns again and again to bear an undesirable child.

What the cost to the State of even one undesirable family may be, is well illustrated by the notorious Jukes family. This family goes back to a Dutch settler in America named Max. Two of his sons married two of six sisters, and the wretched descendants of five of these sisters were traced by Dugdale, who happened to observe that the inhabitants of certain county gaols in 1874 were largely Jukes or relations of the Jukes.

Broadly speaking, the descendants of Ada were prevailingly criminal; those of Belle, without sexual control; those of Effie, paupers. This first investigation of 540 persons of Jukes blood, and 169 related by marriage or cohabitation, many of whom were probably blood-relatives, showed that of 162 marriageable women, 52·4 per cent. were known to be harlots; that of 535 children born 335 were supposed to be legitimate, 106 were illegitimate, and 94 unknown; so that at least 23·5 per cent. of the children were illegitimate. Further, over 20 per cent. of the males and 13 per cent. of the females received outdoor relief, while 13 per cent. of the males and 9·5 per cent. of the females received almshouse relief. Fifty were convicted criminals; fifty were prostitutes, forty of whom spread disease. There were 250 arrests and trials. The direct cost to the State by 1874 was over a million and a quarter dollars.

Forty-two years later Estabrook brought the study of this worthy family up to date. The record was almost exactly the same, though a very slight improvement appeared. This was due to the fact that, instead of living in an isolated mountain district, where inbreeding was close, increased facilities of transport had led to a wider scattering of the family. Thus out of 654 persons who could be roughly classified, only 323 were absolutely valueless; 255 were more or less unskilled labourers; and 76 were actually good citizens. Crime was rather less, harlotry about the same, institutional care rather higher. Admixture of alien bloods had improved the Jukes family a little; but at the expense of the other families, which must have been degenerated to some extent by the undesirable Jukes heritage. Altogether there were about 2000 additional members of the clan, but fortunately the birth-rate had dropped by almost 50 per cent.

Illustrations of this kind could be multiplied. The State of Virginia is at present trying to cope with a lawless and degenerate population in an isolated mountain district. The Hill Folk, the Nam family, the Kalikaks, the Tribe of Ishmael, the Zeros—to give them the sobriquets which hide their identities—afford equally startling statistics. The Zero family is of particular interest, since an attempt was made to reform it by bringing up some of the children in good homes. The experiment failed, for every single child ran away sooner or later and returned to the family habits.

In this connection it is interesting to notice that a special effort has been made in Denmark to develop and train the mentally deficient. In this case there was marked success; but unfortunately it is found that now the second generation is coming along, and it shows the same feeble-mindedness as its parents. Denmark merely managed to propagate the unfit under rather better conditions; and I believe the experiment is now to be discontinued.

We have seen, then, that good qualities run in families; and that in some cases at least there seems to be in- heritance of highly specialised types of ability, such as music or science, as well as of more generalised types of ability, such as government or military command; and we have also seen that criminality, lack of sexual control, and feeble-mindedness are inherited with equal sureness.

Now, though it is certainly true that very many factors must go to the mental make-up of a man, it is equally true that one bad factor may render hundreds of good ones useless. We have already noticed that a double dose of the yellow factor is fatal to a mouse after a short period of development: there is obviously a clash between the functioning of some hormone and the pigmentation factor. We also know that in such disturbances as Graves' disease, or in the toxic delirium of typhoid, the perfectly sound mental mechanism may be prevented from functioning by the presence of the poison. It is quite possible, then, that feeble-mindedness, certain forms of criminality, and so on, may be due to a very few undesirable factors, or even a single one, the presence—or absence—of which renders the rest useless. Feeble-mindedness, epilepsy and chronic criminality, seem to appear as alternative results of some undesirable marriages. The grain of dust on the balance-wheel of the watch stops the whole machine.

In America there are 300,000 insane, feeble-minded, or epileptic through hereditary taint. It has been calculated that this implies that one in fourteen of the population carries the undesirable recessive which, if it meet another, will emerge in the next generation as one of these disastrous conditions. And it must not be forgotten that even the single dose of an undesirable recessive may easily lower the mental and physical vigour of the carrier. We have no reason to postulate the perfect dominance of the desirable factor: all the evidence is against it. The so-called normal offspring in a family exhibiting several cases of mental deficiency are generally of a very low grade.

Fortunately, there is strong reason for believing that the estimate of one in fourteen carrying the factor of mental deficiency is far too high. Sexual selection steps in. A high grade man generally marries a high grade woman; low grade generally marries low grade. Consequently it is probable that far more than one in fourteen of the lower grades carries the undesirable recessive, and far less than this proportion of the high grades carry it. In England matters are, statistically, very much the same as in America, or even a little worse; but owing to the social conditions it is probable that there is even more sexual selection, so that the better stocks in England may be less contaminated than we might fear. Nevertheless the situation is serious enough in itself, and still more serious in view of future possibilities. Although idiocy is not the same as crime, nor epilepsy the same as lack of sexual control; although mental deficiency is not one disease but many; statistics do seem to show that in degenerate stocks these crop up as alternative indications of the undesirable germ-plasm; and it appears that defect in one or two genes is probably responsible for the whole gamut of disaster. It is therefore a matter of supreme importance to discover whether the proportion of defective genes in the population is increasing.

It is disquieting to find that the number of inmates of special schools and asylums is in fact increasing; and that certainly means that, in spite of sexual selection, the defective germ-plasm is slowly spreading over the population, thanks to indirect causes which we shall consider later.

The basic reason for this increase in the actually defective lies in the abolition of natural selection; this also causes an increase in the defective germ-plasm which carries a single dose of the undesirable factors, both directly and through the indirect economic effects.

In the old days selection worked fairly well. It was a hard, roisterous existence, in which the weakly were not patched up, but died off. There was a high birth-rate and a high death-rate, and the weaker stocks did not survive to reproduce, as a rule. Defective types who became criminal were hanged if they did not die of gaol-fever, in the days of hanging for a forty-shilling theft, or any other of the 200 odd crimes which were capital. The method no doubt destroyed a good many enterprising citizens to whom the opportunities of commerce were denied, but it successfully eliminated the mentally deficient who had criminal tendencies. The lunatic also was treated with such brutality that a check was imposed upon his fertility. He was not likely to survive treatment, nor was he favourably regarded when he was at large. By such rough and ready methods were the undesirables weeded out, until the beginning of last century.

All that is now changed. Humanitarian ideas shrink from the callous methods of our forefathers, and medical science preserves the lives which would then have softly and silently vanished away in infancy. This must be so: we cannot go back; nor can we dream of letting that appalling wastage of miserable lives disgrace our civilisation again. But we must face the consequences of our humaner ideals without flinching, or the last state of the nation will be far worse than the first. The shadow of disaster already hangs over us.

Before we turn to this side of our work it will be worth while to see what indications we can find in history of the working of artificial selection in mankind; for the nineteenth century was pre-eminent, not for the industrial development on which the history books lay stress, but for the more subtle and far-reaching substitution of artificial selection for natural selection—and artificial selection on the wrong lines, unfortunately.

When transportation replaced capital punishment, many undesirable citizens were sent to Australia; but with them went many victims of misdirected enterprise, and many who had been driven to crime only by hunger. The violent criminals who became bush-rangers were quickly disposed of, and from the rest grew up a population (affected however by subsequent immigration) enterprising, vigorous, with perhaps an element of lawlessness in their independence; and perhaps also with too little of the ruling germ-plasm to provide the statesmanship necessary for wise and foreseeing legislation. Whether the unpractical characteristics of the Australian town populations, so radically different from the rural populations, belong to the later importations I do not know; but it is reasonable to suppose that they do. The Chartist Riots and other industrial troubles sent many overseas, willingly or unwillingly; later many who failed to make good in England emigrated; and one is inclined to see in the difficulties and the successes of the Commonwealth a reflection of the stocks which formed its backbone.

Three great experiments in colonisation show biological results of a marked description. Spain sent out to South America her youngest, noblest, and most bigoted. They conquered, gained riches—and destroyed their own nation. Women were not sent out, and the men bred with the native races; producing a half-breed nation with some good qualities and some bad, but with no feeling for the old country, and definitely inferior to the original immigrants. Spanish South America has never really gone ahead under its own steam, and Spain never recovered. What went out from Spain never came back. If you take the best from a nation, by whatever means (and in England we are doing it in quite a different way), the nation will never recover.

The second experiment was that started by Richelieu and perfected by Colbert. New France was a failure at first, but when Colbert took control he sent out only good stocks—artisans and so forth—into French Canada; he sent out shiploads of suitable wives; any man evincing anti-social traits was promptly banished from the community. Bachelorhood and spinsterhood were heavily taxed; children encouraged by bonuses. The result is seen in French Canada to-day. Perhaps there is no more stable, hard-working, and homogeneous population in the world.

The third case is North America. These colonies were originally made by people, largely from the British Isles, who hated tyranny, whether of Church or State; people who could not stand the conditions of the day in Europe. To these were added fine families of adventurous souls who went out to make their fortunes and to discover new lands.

The first class tended to be narrow-minded and sometimes cantankerous, but vigorous beyond all others in Europe. You would therefore expect to find in the New England settlements a peculiar fanatical idealism, and a marked lack of cohesion between the larger communities. The early history of the· United States exemplifies these traits unmistakably. There is idealism to the nth degree; idealism which could perpetrate the cruellest of witch-burnings; whose hell was hotter than that of the Inquisition even; and whose religious intolerance was strongly marked. To this day America suffers from an uncontrolled idealism, especially among its women, which goes easily in harness with a commerical enterprise which is ruthless and often amoral. The lack of cohesion shows even in the different aspects of individual lives. New religions, prohibition, gullibility, live side by side with practical genius, terrorism, law-breaking, and a fundamental self-distrust. America is the first country to practise sterilisation of the unfit, and the last country to offer official opposition to the theory of evolution.

But one star differs from another star in glory: the laws of the component States are different. There is no real cohesion, save in the ideal of the United States. Evolutionary doctrine is not illegal in New York; Tennessee does not practise sterilisation of the unfit.

Such is the nation founded on a great race of hardy individualists whom England could ill spare. But later immigrants have done her grievous injury; how much, only the future can show. It seems to be certain that if the old stock had expanded at the natural rate, the population of America would now be as large as it actually is after importing a far less desirable population from southern Europe, the Balkans, and Poland. But that is another story, which may be left untold for the present, and we will simply note that these three examples of colonisation do give distinct indications of biological significance.

Chapter VII

INHERITANCE AND SOCIAL AFFAIRS

We turn from the rather suggestive experiments in colonisation to our previous question: "Is the undesirable germ-plasm increasing in England?" to which we will add another: "Can we hope much from an improved environment?" The people we have to consider are those who are feeble-minded in the sense that they are of rather low grade; we will use the term loosely to cover all those who are well below the mental level of a good citizen, and who at best cannot support themselves except under rather careful guidance. There seems to be no doubt whatever that this class has been increasing by leaps and bounds; and somebody has got to pay for the upkeep of the lunatics and feeble-minded for which it is responsible. A large amount of the money spent on so-called social services goes to asylums, special schools, workhouse and other relief, prisons, refuges, doles to the unemployable, and the like.

No doubt we are more sympathetic than our fore-fathers; but there is a biological basis for the increasing insistence of these demands. Unemployment returns of course depend to a large extent on the conditions of world-trade; but, broadly speaking, when trade is slack the better workmen are retained, the worse are stood down;[1] and at the bottom are many who are unemployed

---

[1] This refers to normal trade-cycles: in the more marked depression which succeeds widespread war and in great industrial changes, some of the better workers are thrown out of work.

because they are really unemployable. They have never been in continuous employment: from boyhood casual labour, if any, has attracted them. They constitute an undesirable section of the population, because, through no fault of their own, their hereditary make-up is less good than that of the rest.

Such a population has to be paid for by those who earn sufficient money, or possess sufficient capital, to be taxed, directly or indirectly. Social services can only be paid for by taxation; and if a man costs the State more than he pays in taxation, he is a burden on the rest of the community. This means that there is less money to support the children of those who pay the taxes: if their money goes in taxes they cannot pay for their own children. The result is an enormous decrease in the birth-rate of the most desirable stocks. The birth-rate of England is sinking: that is in itself a good thing, for various reasons; but when the nature of the fall is examined, satisfaction gives place to grave anxiety. It is in the best stocks that the fall is most marked. Amongst the class from which casual labour is recruited, and amongst those of the lowest mental grade, there is hardly any reduction at all.

In the Middle Ages younger sons of the upper class could only enter two professions, the Church and the Law. Others might, and did, descend in the social scale, and intermarry with the yeoman class, thereby strengthening it. We find many famous old names scattered over the countryside to-day, as Thomas Hardy reminds us.

Later came the Navy and the Army as national services, giving a fresh opportunity to younger sons; and soon certain departments of adventurous commerce offered fresh scope. Families were large, averaging six or seven (now the average in the same class has sunk to under three). Of course the death-rate was high, so that the weak were eliminated.

Then came the period of the industrial expansion and the Napoleonic wars, and a new call for children. The result was an enormous expansion of population.

In the last quarter of the nineteenth century a decrease set in, but entirely in the upper part of the population. Then the birth-rate became most restricted of all in the skilled artisan class: a class with very many desirable qualities, and great vigour. We shall see later that out of the middle classes, rather than out of the upper, genius emerges. Meantime there is no alteration in the birth-rate lower down: the labourer and the feeble-minded are still reproducing at the same rate.

The result, which has been foreseen by biologists for many years, is that there are not enough people of first-rate ability to fill the places at the top, and we have to be content with the second-rate. The Church, the Army, and the Navy have been crying out for more men of the old calibre as officers, though since the War a contraction in the fighting services has slightly masked this demand. Everywhere there is plenty of room at the top. Great captains of industry, great organisers, great servants of the State are all too few, as those who search for men fit to reorganise a group of industries, or to direct finance, know only too well.

The fact is that, partly through luxury, but mainly through the burden of taxation, the most desirable stocks in the nation are not reproducing at anything like the rate necessary to keep up their numbers. America is faced with the same problem. It has been found that with a 90 per cent. marriage-rate, the children of

American men of science are only 1·88 per family, or about seven-tenths of a son per family.

Meanwhile, the classes lowest in the scale go on reproducing at the old rate of nearly seven per family; and the burden grows worse and worse. This class does not rise because it cannot. Even two hundred years ago ability was able to rise, and we have made the social medium far less viscous; yet few escape from the lowest grades into those above.

The fact is, that the unexpected, yet obvious, result of our democratic ideas has been the replacement of a *socially* inferior class by a *biologically* inferior class. This is one of the most important phenomena of the world to-day, and we will spend some little time in examining it.

In the past, as I have pointed out, the yeoman class, for instance, was constantly recruited from the great families; and the labouring class was to some extent recruited both from the higher classes by illegitimacy, and from the yeoman class also by intermarriage. The abler germ-plasm was constantly being added to the less able. It was not easy to rise, but there was built up at the bottom a thoroughly sound population, since the death-rate was high as well as the birth-rate.

Then came the Industrial Revolution. If you look at the outstanding families of the nineteenth century it is striking to find very many of these prominent for ability, prominent for wealth, coming from the yeoman class, more or less. As we have already said, a great number arise from the Yorkshire dalesmen. Water-power replaces hand-weaving, steam replaces water-power; the family moves from the mill to a larger house; then, the fortune made, the old town is left, and the family moves in a higher social sphere. The change is slow, and takes several generations.

Now the rise becomes more rapid. Instead of occupying three or four generations it takes place in one or two. The man who is ambitious waits to marry till he has made his position, and marries late, which means a lessened birth-rate. The social medium is less sticky, and ability rises with extreme ease on the whole: educational facilities, and afterwards every opportunity, come to those who are really able, and they do rise. For the last hundred years this has been going on to an increasing extent.

The net result is that the middle strata are no longer reinforced from the upper, nor the lower from the middle, and the ablest germ-plasm is being constantly removed upwards. In two generations, now, the really able germ-plasm from the lower social strata may be in the House of Lords.

You are left with a bottom stratum which is not worth very much, biologically speaking. But this is not the worst of it.

We can see the beginnings of a still more fatal process taking place even in the Industrial Revolution. What happened then was that the abler, more vigorous of the country population migrated to the more adventurous life of the towns. This had two effects from the biological point of view. In the first place, rigorous natural selection set in. Many died of tuberculosis, and various urban diseases. Many, excellently suited to country life, could not stand the conditions of town life, and died off. But, further, urbanisation is sterilisation, owing to economic conditions. When child-labour was allowed in the factories this was not so evident, but, as we shall see, as soon as this ceased, families became much smaller.

On the countryside this migration also produced biological results. What was left in the country was the less vigorous, the less independent and adventurous type; and the result was the decay of village life. This last is by no means due to a decay in agriculture; rather it is a contributory cause of that decay. In the old days, when transport was difficult, there was much inbreeding in the country villages; but on the whole the sound stocks interbred with each other, and the unsound with each other. Consequently the village idiot was common enough to be proverbial; but the able man was also common; and it was the able man who was taken away. In village decay we see on a small scale the same experiment that we are trying on the nation under the aegis of democracy; and though a further increase in transport facilities is resulting in a slight revival of village life, there is little hope of that for the nation.

We have already noticed the sterilisation of the upper classes, and hinted at its cause; this we must study later in more detail.

Much evidence, such as the records of maternity benefit paid by the Friendly Societies, goes to show that there is at least as serious a sterilisation of the thrifty artisan class. And we have further noticed that there is no such sterilisation of the layers of casual labour—labour which is casual because it is not fit to be anything else. In many cases it is mentally deficient in the sense of being inherently unqualified to pass beyond the level of the elementary school, to write a letter, or to read a newspaper with any understanding. It is calculated that 25 per cent. of the population of America falls into this class; and there is no reason to suppose that England is much better off. It will be seen that I am here stretching the term mentally deficient to cover cases where all the genes, or a large number of them, seem to be of low quality. The true mental-deficient probably lacks a factor, whose absence renders useless the good factors which he possesses. Perhaps it would be better to use the term mentally inefficient to cover both classes.

Clearly what is happening is that we are trying an experiment in the segregation of hereditary factors on a large scale. But the disastrous side of the experiment lies in the fact that, having segregated the desirable germ-plasm and brought it to the top we proceed to sterilise it, while we leave the lower germ-plasm, from which we are steadily draining all that is good as it appears, with an unchanged birth-rate. We are breeding the un-desirable part of the population, and making the desirable part pay for it instead of having children of its own. It is impossible to exaggerate the gravity of such a situation.

History furnishes us with parallel cases. Though no doubt the spread of malaria had a good deal to do with the fall of Rome, and very likely of Greece also, it was not the prime cause. Even in the days of Augustus 1000 sesterces were given for a child of Roman parentage in order to encourage breeding. Even by the time of Trajan they had to go to Spain for their Emperor, and thereafter almost all the Emperors came from Africa, Syria or Gaul. The old noble families were gone. Wealth and sterility at the top; free games, free food, and free breeding at the bottom; slaves, freed by the thousand, forming a helot class; and a cry going through Rome that there are no men to fight. The army is made up of Gauls and other races.

Rome fell. The germ-plasm that made Rome great no longer existed.

In Greece we find exactly the same thing. A wealthy upper class comes into being, and is self-sterilised. A large helot class comes into being, and there are some slaves of ability who are petted and pampered; but theirs is not the germ-plasm that made Athens what it was at its zenith. Probably the ability of the ruling Athenian families has never since been equalled in any population of the world. The Persian wars certainly accounted for much, but there is no doubt that the complaint of Polybius that his country was going down in numbers in the second century B.C. was true; and by the end of the first century A.D. Plutarch states that in the whole of Greece could not be found the 3000 soldiers which little Megara sent to the battle of Salamis. But long before this the great germ-plasm of Greece was dying out. The days of Pericles have gone irrevocably; Greece of to-day is not populated by the seed of those men who made the great age. And Greece fell. It fell mainly because the upper classes failed to breed.

Mutatis mutandis the story is the same as the story of the fall of Spain, whom South America robbed of her best blood.

Almost in our own day we can see the same thing in the State of Virginia. The old colony was the foremost and finest in all North America. During the Civil War the first people to come forward were the old Virginian families. They came in first and fought longest, and they had the highest death-rate. The greatness of Virginia was destroyed.

Both on a small and large scale, then, we see the fatal effects of extinguishing the ablest of a race; and it is exactly this which is now taking place throughout the civilised world. In every country the cry goes up, "Why have we not better men at the top?". The answer is simple: we have been breeding them out of the race. The bottom of every profession is crowded, the top is empty. It is easy to rise, and people of very moderate ability are rising from one class to another.

Even if the man who rises is of first-rate ability, he suffers the same disastrous influence. He tends to marry late, because he is going to marry out of his own class; and that means again a partial sterility, in the sense of fewer children. The nineteenth century was far sounder than the twentieth in this respect as we have seen: then, if a small bricklayer became a master builder, and his son a great contractor, and *his* son or grandson a peer, disaster did not follow. The rise was slow enough for each generation to marry early, in or only just above its own class. The sterilisation of a rapid rise is far more marked, so far as this factor is concerned.

Now it is interesting to notice that practically all genius comes from the middle classes. Havelock Ellis studied British genius, and found that a comparatively small proportion of the persons of the very highest ability arose among those whose position was obviously affected by the accident of high birth, so far as actual numbers were concerned, though the percentage of eminence and genius might be even fifty times as high as that in the normal population. But the middle class is so immensely larger that in actual numbers it supplies by far the majority of the geniuses. The ranks of unskilled labour, large as they are, produce an almost negligible percentage, and even skilled labour produces less than 10 per cent. —a striking fact when one remembers the relative size of the classes.

The upper class are far from homozygous, of course,

but they are more or less segregated as far as certain desirable qualities are concerned: those qualities which make the proconsul or the public servant. In the middle class there is still a vast amount of valuable germ-plasm; and good material is being added from below. From the nature of the case it is heterozygous; and on the whole you are more likely to get heterosis vigour from cross-breeding in the middle class, due to the happy combination of factors, than you are in the more segregated upper class, which has probably achieved the best of which it is capable in some cases, and is a very small minority; or in the lower class, from which the valuable qualities have been steadily removed through the rise of their possessors into another social layer. The whole matter grows clear as we look at it with the knowledge derived from our study of inbreeding and outbreeding.

The fruit of democracy is biological segregation. Its results have been and must be the replacement of socially different layers by biologically different layers: the formation of a helot class of lower intellectual and practical ability: people at the bottom who must always be at the bottom because they are biologically inferior. This is a situation which no one can contemplate with equanimity, since many of this class will never be anything but a burden on the community. Of the political causes and effects of the business we shall speak later.

But when the realisation dawns that we are in fact actually *selecting* this class at the expense of the abler: the upper and artisan classes sterilised, the lowest unaffected in birth-rate; the mind stands appalled. There is no question that democracy hitherto has been selecting the undesirable, and making the desirable pay for them, so that those whose value to the nation is highest cannot afford to have families, since they have to pay for the economically unfit and at the same time face a rising standard of living all round.[1]

> Science finds out ingenious ways to kill
> Strong men, and keep alive the weak and ill—
> That these a sickly progeny may breed,
> Too poor to tax, too numerous to feed.

War and humanitarianism have effects serious enough; but we are not even content with those. Fifteen years ago one would have said, and did say, that it looked as if our number was up. But more cheering elements are possibly beginning to appear.

It is true that we have countered the natural selection of past years by our increased medical knowledge and our growing humanitarianism; but have not faced the responsibility imposed upon us by this development, nor adopted a sensible artificial selection. We have said, "Yes, of course you must save the weak and ill and do the best you can for them"; but we have lifted up hands of holy horror at the idea that we ought to see that the weak and ill do not breed more rapidly than is desirable. We have piously turned up the whites of our eyes at the idea that, if you know enough to interfere with the course of nature, you have got to take the responsibility of your knowledge, and supply in its place a selection which is based on knowledge also. We murmur "Liberty of the individual!" and pass by on the other side, leaving England to die. The absolute density of the population cannot go beyond a certain point:[2] natural

---

[1] The cost of education of a normal child is £12, of a mentally deficient child £93, I believe.

[2] Probably this is not a *fixed* point, it is likely to vary from age to age with the conditions of civilisation and invention.

selection saw to it that this was achieved by eliminating the least fit. We have substituted an artificial
selection which achieves its object by eliminating the
fittest.

But the matter is already to some extent righting itself,
in rather ignoble fashion; instead of righting itself in the
upright and moral way one would wish to see, but for
which we are not men enough. We have encouraged the
ideal of luxury and comfort (probably a good and right
thing, though it can be over-done); and this ideal is
penetrating to the lower levels. Selfish comfort makes
for a lower birth-rate, and there is beginning to be a
lower birth-rate farther down the social scale. This has
not gone far, and will not, among the very lowest levels,
and there lies the most serious menace; but there is no
question that sooner or later—and the sooner the better
—sterilisation of the really unfit will have to be practised.
To allow the feeble-minded to breed is a crime against
civilisation, against humanity, and against morals. Let
them enjoy as happy a life as possible, but in the name
of humanity sterilise!

But even then a danger lies in the motive of luxury
which would not be present if we exercised control on
higher grounds. Luxury may lead to over-reduction of
population. Indeed I am far from certain that this will
not be a real problem of the twenty-first century. Too
small a population is as great a menace to peace as one
that is too large, or one that is inferior in quality.

It is often objected that sterilisation will not eliminate
feeble-mindedness. This is perfectly true. But if for the
sake of argument we regard feeble-mindedness as being
due to a double dose of a recessive factor (and this,
though a simplification, seems to be the right kind of

explanation), by sterilising all those who show the double dose we should prevent the contamination of the race by the devitalising single doses which result from the marriage of a feeble-minded with a normal, as well as rendering the number of people in a generation showing the double dose, either due to the mating of two parents carrying the single dose, or to the mating of feeble-minded parents, progressively less. You can never eliminate a recessive; but you can encourage it, as we have been doing, or discourage it.

You will run little risk of destroying genius by sterilising the lunatic and the imbecile; for in spite of uninformed opinion there is no relation at all between genius and neuropathic taint, as Havelock Ellis has conclusively shown (where genius and instability are allied in an individual they can generally be shown to have come in from different ancestors); and it is possible to deal with the clear cases before bothering our heads about the borderline ones.

At present we simply execute temporary repairs, and then turn out the unfit to breed unchecked. Sterilisation, which need in no way affect their lives, would prevent this without inflicting the slightest hardship. We talk sentimentally about the liberty of the subject, without even realising that many of the wretched women we now support in workhouses cannot ever be allowed to go out, because they have no self-control. They must be, and are, kept virtual prisoners.

The cheap appeal to the liberty of the subject will not bear inspection for a moment. The mentally-deficient offspring have to be kept under control. Moreover if we make it a crime to kill oneself or another, why should we be free to inflict the misery of a stunted existence on those who need never have been born? Compulsory sterilisation is almost certainly undesirable; the association of a penal idea with sterilisation would be disastrous. I would never advocate anything beyond advising those who are brought up in special schools and similar institutions, those who have given birth to mentally deficient children and have reason to suspect a genetic taint, and those who desire to marry into stocks where the risk of mentally deficient offspring is great, to be sterilised; or, where the afflicted person is incapable of forming a judgment, advising the parents or guardians in the same sense. The strictest confidence should be observed, as in other medical matters. I believe that the response would be satisfactory, especially if it was made clear that where the risk of undesirable children was great institutional care was the only alternative.

Experience might add other reasons for giving such advice; but I doubt if it could ever justify us in going beyond the offer of these two alternatives. The idea of compulsion is repellent; the practice would be difficult, and might easily result in failure.

Unless you have a sound germ-plasm you cannot have a sound nation. That is the aspect of the matter that forces itself upon the biologist. It is no good burying our heads in the sand, and hoping that Nemesis will not discover us.

But mere luxury is beginning to do what thought has neglected; there is hope of a reduced birth-rate in all but the least fit of all; and there are at last some signs that those at the top are beginning to wake up to their own responsibility. We must not forget that the breeding of the fit is as important as the elimination of the unfit;

and that our duty to our neighbour is a duty to the unborn as well as to the born.[1]

We may now turn to politics.

Democracy is based upon two biological ideas which are certainly contradictory and both of which are probably false. The first is that all men are potentially equal; the second that you can make silk purses out of sows' ears. The second proposition is, of course, whole-hog Lamarckianism.

The first need not detain us. Its falsity is evidenced by all that has gone before. Hardly anyone would be so bold as to defend it, yet tacitly it inspires many political utterances and ideas. Merely to study the effect upon the descendants of a marriage with an unstable person, or the effect upon fertility of marriage with an heiress, offers evidence enough. Intelligence tests and specialised abilities give overwhelming denial to such a view.

The second proposition deserves a little more consideration. One of the root ideas of democracy is that by improving the environment you can improve the individual. This is a peculiarly specious doctrine, for it is in a great measure true. By improved educational opportunities, for instance, you can raise the standard of education. But how far? That is the crux. Can you actually raise the mental level of the nation; or can you

---

[1] I cannot refrain from quoting as nearly as I can some fine words spoken at a recent private meeting by Dr R. A. Fisher. "The moral basis for coöperation lies in the extension of one's idea of one's neighbour to future generations. We can do nothing for the dead, little for the aged, much for the children, most of all for those not yet conceived. For children we can improve the environment, enabling them to make the best of their innate qualities; for the unborn we can improve the innate qualities. The genetic endowments of the unborn are our acute moral concern."

only develop its component individuals up to the fixed limits of their inherent capacity? Democracy hopes the former: there is every reason to believe that the latter is true.

Two Philippine twin girls were separated at birth, brought up under totally different conditions, and worked in different hemispheres for some time; yet when they were adults it was found that intelligence tests gave practically identical quotients; and that they were still extraordinarily alike in every way. Is it possible to have clearer evidence that education is nothing in the balance when weighed against heredity? Environment cannot do all that optimists hope: Nature is of far greater importance than nurture.[1]

The eyesight of Edinburgh school-children was tested. Some lived in dark and dirty closes, others in healthy surroundings. There was no correlation whatever between the conditions of living and the goodness of the eyesight; but there was the usual correlation between the eyesight of parents and children. Heredity was all-important: environment did not have the slightest effect.

Of course what the biologist calls epigenetic factors are of vast importance. Thanks to books, the learning of each generation is stored up and handed on, so that the next generation can go farther. But this is not the point. Suppose we go farther and say that it is just conceivable that mental aptitudes learned by one generation are passed on to the next? The frequent passage of an impulse across a synapse in the brain in successive generations might pave the way for an easier passage in future generations; though it is hard to conceive the method by which this Lamarckian effect could

[1] Galton's *History of Twins* (*Inquiries into Human Faculty*) is interesting.

be produced. Is there any evidence that such a process does actually occur? We have to answer, None whatever that can carry conviction to the serious student. Superficial observation does perhaps seem to show that ideas which are difficult to one generation come easily to the next—but there is no proof that heredity is really responsible for this; and there is some evidence in modern psychology for the inheritance of peculiar and identical phobias, which may, or may not, be acquired characters —and probably they are not, in the strict sense. Such arguments are highly disputable, and the objections to their acceptance are at present almost insuperable. Could it ever be proved that such inheritance is possible, the hands of the believers in the power of education would be strengthened beyond measure; and it must be conceded that on such lines the inheritance of instincts, such as those of an untrained gun-dog, are most easily explained. But we have already seen how insecure is all the evidence, and how weighty are the arguments on the other side.

However this may be, we must admit two things: that all the garnered wisdom of the years does constitute an epigenetic factor of immeasurable importance to civilisation; and that it still remains true that you cannot make a silk purse out of a sow's ear. All the education in the world will not help the synapses of a mentally deficient child to function as efficiently as those of a normal person; neither will it convert a mediocre person into a genius. The most that education can do is to develop to the full the possibilities provided by the hereditary make-up of the individual, even if we go so far as to admit that those possibilities may conceivably include more than the present evidence justifies us in believing. Yet underlying the whole of our democratic legislation is the assumption that, given opportunity, the under-dog is as valuable to the State as the top-dog. Would that it were so! but it undoubtedly is not the case.

Present-day democracy is the most ghastly blunder that the human race has ever made, as I read the evidence. I can only give my own views, together with some of the data on which they are based. It is far from my wish that a mere dictum should be accepted: I may be wholly mistaken. What I want is that interest should be aroused and that people should think and investigate for themselves; because upon the question of the inheritance of acquired characters, and the proportion of good germ-plasm in the race, depends the future of the civilised world; and the matter is very pressing in England.

Those who have followed my argument must realise that I am bound to utter a diatribe against the present democracy; but I must beg them to believe that I do not advocate the remedy of a return to mediaeval conditions. It is not against democracy itself, but against mistaken democratic ideas and methods, that my argument is directed, as will be evident later. But I do contend that this rash, fierce blaze of legislative riot cannot last; and I do hold that England is leased out like a tenement or pelting farm to most undesirable tenants; and is bound in with inky blots and rotten parchment bonds which hamper her at every turn. We cannot go back, and we would not if we could; but unless we go more wisely forward, to a democracy as yet hardly dreamed of, the fate of England is sealed.

> Well might we ask what Beauty ever coud liv or thrive
> in our crowded democracy under governance
> of such politic fancy as a farmer would show
> who cultivated weeds in hope of a good harvest.
>
> R. Bridges, *The Testament of Beauty*.

Let us review what we have said. First, the birth-rate is definitely decreased, and decreasing. That, in itself is actually a good, and indeed a vitally necessary thing. The second point is, that it has decreased most in the sections of the population which are most desirable, less in the population which is only just able to keep its head above water; while in the grade of the unemployable the birth-rate remains the same or practically the same as of old, while the death-rate has decreased. We are therefore selecting the unfit in exactly the same sense as a dog-breeder who consistently drowns his best pups and breeds from the worst. In the days when there were too many of the top classes, the surplus, and the less fit, went to reinforce the germ-plasm lower down, and a high death-rate weeded out the weaklings of all classes. Now each level must be recruited from the level beneath, where the selective effort of Nature has been already discounted by medical care. Steady selection of the unfit is the net result.

It is interesting to see how this has been brought about. It is not merely a matter of greater medical skill, or even of growing humanitarianism. These are of course reflected in many of the legislative enactments which have contributed to the rise of the phenomenon. Selfishness and luxury have played their part also. But the largest part has been played by the crass ignorance of biological elements displayed by the well-meaning legislator.

Obsessed by half-understood ideals, and unable to see beyond his own nose, he brings forward scheme after scheme for the amelioration of the lot of the under-dog, without having the slightest idea that it will have any selective effect at all; or that it will, in the long run, intensify just those evils that he is trying to cure.

The most pressing need of the day is that politicians should be taught biology. Few politicians look more than ten years ahead at best: the unit of the biologist's thinking is a generation; and it is not until the F_2 generation that segregation begins. The biologist cares as much for the amelioration of the human lot as the wildest fanatic; but he knows enough to realise that such amelioration will not come through the selection of the elements of the race which have shown themselves least fit to cope with the conditions of life.

The politician is surprised and appalled by a growing expenditure on unemployment and on all the aids to the physically and mentally sub-normal. As long as the increase is steady—that is, in the periods when there ought to be a decrease—he merely pats himself on the back over the efficiency of social services and the bowels of compassion of whatever party happens to be in power. When world-events are adverse, and the burden grows suddenly intolerable, he looks back, observes what has been happening, and is appalled. But even now he does not grasp the real reason, he tinkers despairingly with temporary remedies, and says that the trouble is due entirely to world-conditions at the moment, and will pass. He does not know that the term "world-conditions" really explains nothing whatever. World-conditions determined the drift westwards of wave after wave of the races that built Europe; but population pressure, and the shifting of a rain-belt really started the drift.

The biologist has known for the last thirty or forty years that we were going to suffer from an increase of unemployables, and a lack of leaders. The War has terribly intensified the position, owing to the reckless wastage of the best types in the first two years, and the £2,000,000,000, or whatever the sum is, that was converted into carbon dioxide and water, necessarily accentuates the economic stress; but the fact that the condition was bound to arise before very long was clear by 1900, at any rate, and was a common topic of discussion among biologists long before 1910.

We may illustrate the matter by glancing at a few of the legislative acts which have had an obvious biological effect.

In the eighteenth and nineteenth centuries, before the Factory Acts, large families were an asset. Though what each child earned was little (children of five and six might be working as much as thirteen hours a day in the mills, for 5s. a week), the total effect on the family budget was large. The abolition of child-labour turned the child from an asset into a burden during its early years. For the moment we are not concerned with right and wrong, but merely with biological matters; but no one can doubt that a system which held child-life cheap, and broke many children in body and mind, is at least as wrong as any system which breaks adults in body and mind. It had to go. But in going it reduced the birth-rate, among the thrifty especially.

The result was that in about another generation or so the need for an Old Age Pensions Act began to emerge.

When families were large, it was not difficult for the many children to find enough shillings a week to keep the old parents; but as families grew smaller the burden upon the few children was too great. They could not support their own families and at the same time pay much to the old people. The passage of the Old Age Pensions Act produced yet another drop in the birth-rate in just the most valuable class of the wage-earners—the thrifty who were in steady employment and were tending to rise—since children were no longer needed for the support of aged parents.

The enfranchisement of women, and the accompanying economic emancipation which threw open many opportunities to women, again reduced the birth-rate of the wage-earning class, both by delaying marriage, and by the wish that the married woman should still earn, and have the more exciting life of the wage-earner.

On the other hand the thriftless were practically unaffected by these measures. They had never looked ahead, nor did they do so now. Education Acts relieved them of even the small expense of the National School; if there was not sufficient food, the children were fed at school; boots were provided; medical services were given free; unemployment was met by unemployment pay; the sickly and feeble-minded were taken away and cared for. Even such reasons for thrift as used to exist were gone.[1]

Naturally their birth-rate did not decrease. The thrifty felt it beneath them to send their children hungry to school, or badly clothed; these did not.

The increasing cost of such measures fell most heavily upon the middle and upper classes; whose birth-rate promptly showed a rapid decline to half its normal value. In another generation their birth-rate had fallen still lower; the War had reduced by a large percentage those who should have helped to pay for these costly experi-

---

[1] Recently the second mate of a coasting steamer told me that he sometimes felt he was a fool to go on working, as by becoming unemployed he would receive 4s. a week more pay, and maternity-benefit as well when his expected first child arrived.

ments; all the steady wage-earners had small families; taxes and death-duties led to the disappearance of many families from the estates of the countryside, and exercised a depressing influence on country life; but the breeding-rate of the least employable and the feeble-minded had not appreciably decreased. The burden of carrying an undesirable population grew too great; there was no money to spend, so that even the employable began to be out of work in large numbers. But, thanks to en-franchisement measures, the voting-power of the class which benefited most from the State expenditure was made overwhelming, and the "dole", being on such a scale as to provide comparative comfort without work if the family is large,[1] definitely encouraged the breeding of that portion of the population which is usually un-employed. Ignorance of biological consequences can indeed make bad mistakes! The accident of post-war conditions was disagreeably intensified; and the country is within measurable distance of financial disaster.

That is how we stand to-day. Yet there are plenty who say that it is a mere matter of world-trade: who are too ignorant to see that we have been leading up to the situation for the last fifty years by selecting the unfit and discarding the fit.

Paradoxically enough, it seems likely that a Labour Government will wake up to the truth sooner than either of the other parties. The skilled labour of England is

[1] Any *flat-rate* family allowance must necessarily have an effect which is biologically bad, for while the allowance may be suffi-cient to counteract the expense of children in the lowest social grades, it will do little to help those whose standard of living is higher. Consequently the lowest grades will reproduce freely, while the higher ones are unable to do so. And the lowest grades are increasingly inferior in biological value.

beginning to feel the pinch now. As comfort spreads downwards there will be a decrease of the birth-rate lower down. The sterilisation of the unfit, when it is undertaken, will remove to a considerable extent the source of the unemployables; and if conditions at the top become slightly easier the better strains will begin to breed a little more freely. Of this there are, I think, some indications already. (See note on p. 210.)

Unfortunately an awakening is delayed by the fact that the poorer classes are only taxed indirectly. An extra penny or two on tobacco and beer causes grumbling, no doubt; but the untrained mind does not realise that this is a contribution to the cost of social services. The rent goes up; but the wage-earner only complains of a grasping landlord, not realising that rates are included in his rent, or that the cost of repairs has increased.

We see, then, that the trouble arises more from the unforeseen results of social legislation than from the ideals which lie behind that legislation. Any biologist could have foretold, for example, that the enfranchisement of women, however desirable in itself, would have a sterilising effect. Indeed one would have thought that the result would be clear to everybody. It is perfectly obvious that if the attempt to rise in the social scale leads to the partial sterilisation of the most desirable males, the same sort of opportunities will also lead to the same result in the females—and the maternal inheritance is as important as the paternal.

But, as we shall see later, the other side must not be forgotten. A great lowering of the birth-rate is absolutely necessary if the sounder ideals of democracy are to be realised; for the only possible alternative is reduction of population by famine, revolution and war. The danger lies, not in the lowering *per se*, but in the fact of a *differential* lowering, unfavourable to the best, favourable to the worst.

All the legislative enactments of which we have been speaking, and in some cases the very necessity for them, are really based on the biological assumption that by improving the environment you will improve the race: an assumption made in the face of the fact that there is no evidence whatever that acquired characters play any considerable part in the development of a race, and a vast amount of evidence that hereditary characters do.

All that we can reasonably hope from a good environment is that it will give the germ-plasm opportunity for development; and our business is to supply the good germ-plasm first, and then to give it the best environment that the world can possibly provide. By neglecting this first essential, modern democracy, living in a fool's paradise, has done more harm than the cruellest autocracy that ever disgraced the planet.

But it is even yet not too late. Let us theorise for a few moments. It is possible that out of the ashes of a dead democracy will arise a phœnix called aristodemocracy which thinks ahead, allows the sterilised unfit as free and happy a life as their mental level makes possible, encourages the mating and reproduction of the best stocks, and yet stabilises the numerical strength of the population at the level which is economically best by a heavy taxation on families which are too large. Such a state would have perhaps to be founded on a three-generation basis, for segregation takes place in F_2. The social medium must be fluid, but not too fluid. Wealth and position won by merit must not revert immediately to the State on the death of the person who won it; but the next generation might pay very small death-duties on the inherited capital which it received, just sufficient to pay expenses of administration, and later generations, the F_2 and F_3, might pay at a higher rate on the remains of this capital. Should the second generation prove socially undesirable, however, the whole of its inherited wealth should revert to the State. It ought not to be beyond the capacity of parliamentary draughtsmen to invent a law of entail based upon three generations, or beyond the ingenuity of legislators to invent some means of transferring the capital of those frequently guilty of antisocial offences, and of those who are imbecile or insane, to the State; or even of compelling those who beget imbecile children to pay something really considerable towards their maintenance as well as ensuring that they do not beget any more.

After all, by far the most momentous event in the whole secular history of the world characterises the last century—indeed the last half-century. During that period all civilised races have worked to destroy natural selection and replace it by artificial. But they have not understood the responsibilities of their discoveries, and have applied their artificial selection at the wrong end. The Russian method of destroying the aristocracy (perhaps in that case rather a poor stock) and with it the middle classes from which the driving power of a nation is replenished, is the logical *reductio ad absurdum* of the method. Applied widely it would in two generations destroy civilisation, without leaving sufficiently good germ-plasm to make possible the rebuilding of any civilisation worth having for many generations, if ever. But England and other countries are really doing the same thing in a less violent fashion; and something or other is bound to happen before long as a consequence. It might even be legislation of a biologically sane description.

There is no limit to the possibilities of a biologically wise legislation which is sufficiently moral to face facts, and to realise that the perpetuation of an unsound stock is far more wrong than a compulsory operation on a person, which is, in the male, hardly more serious than having a boil lanced, and in the female no worse than having a healthy appendix removed. We would have no interference with marriage in our aristodemocracy; but a medical board with a court of appeal would examine the bride and bridegroom, and if the stock or condition of either were unsatisfactory, a slight operation would prevent the possibility of the tainted stock being perpetuated through that person, though no bar whatever would be opposed to the marriage. Even if we have not quite reached the simplicity of surgical treatment implied, we are not far from it.

I think we should decide to make enfranchisement depend on economic efficiency, and have an educational qualification for extra votes.

There should be little real difficulty in selecting the children who are likely to do well in a particular line of life: already such tests are being applied by industrial psychologists on a small scale. The really brilliant child is almost always marked out from a very early stage, and it is about fifty times more likely (judging by Cattell's statistics of scientific men) that he will come from a family showing marked ability than that he will emerge straight from an undistinguished ancestry. The brilliant geniuses of the world have been conspicuous from infancy: they can read and write by four, know Latin by six or seven, and Greek almost as soon, master abstruse mathematics by eleven, and by twelve or thirteen are surpassing the ordinary educated adult. In the very highest class, judging by childish performance, come the philosophers—an interesting answer to the people who, unable to understand any philosophy themselves, write it down as rubbish. And the Newtons, Laplaces, Davys and Youngs make discoveries that revolutionise thought almost before they have come of age, or at any rate in the early twenties.

In our aristodemocracy we shall have to be on the look-out for ability; and we must devise some method of not expending most of our energies in making the dull scholar into a mediocrity while we keep back the child who could go at twice the pace, as we are inclined to do at present.

It is a fact that in America only 1 per cent. of the children have intelligence quotients above 130, where the norm is 100 and the feeble-minded are below 70; and there is no reason to suppose that we are better off. Yet it is from this 1 per cent. that practically all advance, all government, and all wealth will come. Our aristodemocracy must begin with this 1 per cent. and not waste it, but gradually make it into 2 per cent. and then 3 per cent.

But all this is too easy and simple as it stands; and if it were not, the world is not ready even to wonder whether there is anything in such dreams of a saner, cleaner society. When the young begin to see visions instead of leaving it to their elders to dream ridiculous dreams it will be time enough to talk of practical details; for when that time comes things will begin to move. There is power in a vision, but none in a dream. Therefore we will leave the foolish dream and return to everyday facts.

None the less, although a mere dream is not a reasoned scheme, legislation with a sound biological background must come in the near future, or be too late. Blindness, prejudice and an obscurantism which hides sometimes mere ignorance, sometimes an unconscious psychological twist, sometimes a more or less conscious will to power, under the cloak of religion, have yet to be overcome if the present civilisation of the white races is to be saved.[1] Perhaps the balance of probability still tilts on the side of the old, blind methods of past history: the decline and fall of a civilisation, a period of chaos, and then again a slow rebuilding. But it is difficult to see to-day any race fit to rebuild; and yet we know the races of the world as they were not known when the great waves of conquest rolled over decaying ancient civilisations.

We have spent some time in seeing how mistaken legislation can have unexpected effects, and lower the stamina of the race. We must now consider the size of the population itself.

In old days this was regulated by war, famine, and disease—strongly selective in their functioning. War, being in the Middle Ages mainly a professional occupation for foreign mercenaries, did not kill off the strongest in a given nation as much as might be expected, though it was fairly indiscriminate as regards the conquered country. And even where the nationals were fighting, on the whole the strongest tended to survive, for disease and exhaustion killed off more than battle; and in a hand-to-hand encounter the stronger man usually slew his opponent.

---

[1] The biological effects of the *Ne temere* decree taken in conjunction with more recent Papal utterances are obvious and suggestive.

Modern war exactly reverses this. In England during the Great War, until compulsion was introduced, the finest stocks rushed to the front, and were wiped out by the indiscriminate action of high explosives which give no more chance to the strong than to the weak. Such diseases as typhoid were practically non-existent; and the one serious epidemic, Spanish influenza, killed indiscriminately, since none of the population seemed to be immune. Although the women of the best stocks were left, there can be no doubt that our recent experience has proved modern warfare to be dysgenic in a way that the older kind was not.

Famine we have for the time conquered; though it may be only for the time. But the probabilities are that any famine likely to occur will be tided over, if not with imported then with synthetic foods—any catastrophic famine, that is to say. General food shortage we shall consider immediately.

We seem in a fair way to master the more selective diseases; though Nature will no doubt invent others, by the variation of some bacterium, that may take a heavy toll before the counter-measures can be worked out and the disease controlled. But taking it all round, we have successfully removed the old checks to population. We have also legislated for the person who is unable to control himself, whether in regard to alcohol or drugs or sex or other things (and America has done it more than we have) in order that his germ-plasm may survive. Altogether, we have set ourselves to make a pretty mess of Nature's safeguards.

Such measures, preceded by the call for children uttered as the combined effect of the Napoleonic struggle and the Industrial Revolution began to be felt, produced a vast increase of the population in the nine-teenth century. As population began to outrun means of subsistence—for various reasons, from potato-famine in Ireland, and the corn-laws which synchronised with the mechanisation of industry in England up to the "hungry forties", to the change of European and American nations from consumers to producers of manu-factured articles towards the end of the century—the obvious remedy seemed to be emigration.

Now there is a fallacy in the idea of remedying over-population by emigration. A country will fill the vacancies caused by emigration in a very few years: the relief is purely temporary. Of course, if you emigrate your undesirables and keep your good stock, you will score heavily; but anyhow your population will quickly return to saturation-point. When that saturation-point is reached it will become more or less stable. While the earth was fairly empty, migration was the remedy for over-population due to a shifting rain-belt or a sudden racial step-forward, though even then smaller and more backward populations had to be destroyed or absorbed; emigration is but a useless vestige of that old instinct. If you cannot migrate, you must over-crowd to the limit of a rising standard of living; and then things are likely to happen, as Europe has lately found to its cost.

From the point of view of the country to which the emigration takes place (which can only be one which is developing and under-populated) the effect is interesting. America affords the most striking example. Before the middle of the last century the immigrants were of a vigorous type—English, Scotch, German, Scandinavian —which showed originality and ability: sound stocks seeking to escape first the tyranny of Church and State, and second the tyranny of industrial conditions. Then came a large Irish immigration, of markedly lower worth on the whole, but by no means confined to the derelict strata. Since about 1882 over two-thirds of the immigrants have come from the flotsam and jetsam of south Europe, the Balkans, Poland and Russia. In 1897 the proportion of immigrants from these countries was 87 per cent. Their intelligence quotients are low: very definitely lower than the average of the whole population, even though the really feeble-minded are turned back at the ports. We are reminded of the fact that recessive feeble-mindedness can be carried without any obvious symptoms, for mental defect is increasing rapidly in America.

Such immigrants tend to produce cheap labour, and it was some time before America realised that this was not a good thing economically. It tends to restrict real wages; for a minimum wage becomes necessary; and people begin to think in terms of a minimum wage.[1] Then comes along a practical man (who is hailed as the greatest of altruists when he is only the man with the best business head) who sees that if the standard wage is $3 a day, and the best workmen are really worth $8 a day, he can offer $5 wages and take his pick, to his own vast profit. His factory will be the most efficient in the world. But there are few workmen who are worth the $8; so it will not be possible for many factories to follow his example successfully. There are not enough first-rate men to supply more than a very few factories with such material. But the wages of the others tend to

---

[1] It is worth considering whether the demand for a minimum wage in England is not largely caused by our increasing production of the less efficient type in the population.

be forced up to an unproductive level in sympathy: un-productive, simply because the men are not worth the wages. Meanwhile the workmen who are only worth $2 a day must be employed at the standard rate; and in order to maintain them the production of those who work with them has to be lowered to their level.

Now the low-class intellects which have been squeezed out of Europe have made up the bulk of American im-migration since 1882. No longer is America getting the most vigorous blood of Europe; her age of expansion is ending, and the immigrants are pulling down the whole standard by tending to reduce wages and thus to force the demand for a minimum wage, which may easily become uneconomic if the standard of production is lowered, to increase taxation, and to add to unemployment. They want looking after in one way or another, whereas the high-grade man is naturally thrifty and self-respecting, and resents a grandmotherly interference.

But there is worse to come. By importing these large and valueless populations you sterilise the population that is already present. The reason is perfectly simple: that population is taxed to supply the needs of the im-ported population, both directly and, still more, in-directly; and the home population, which is of much higher value, has fewer children. It is apparently a solid fact that the population of America would be the same as it is now, and of a much higher grade, if there had been no immigration after 1820. Professor East's study of this problem of immigration in *Heredity and Human Affairs* is well worth reading. The other objections—that you have German colonies, Italian colonies, Russian colonies, Czecho-Slovak colonies, Greek colonies, Polish colonies, and so on, preventing any real national cohesion, and importing their own brand of crime—are obvious. The tragedy of the United States is, perhaps, that there is no American nation; it makes for a radical self-distrust which sometimes tries to hide itself under a show of "boosting".

Of the effect of cross-breeding on all these nations we as yet know little; but our next study will be concerned with the inbreeding and outbreeding of nations, and we shall see reason for grave doubts. At present we can only say of the export of the world's undesirable stocks to America, that however nice it may be for their parent-countries, it is very hard on their adopted country.

However America has waked up to the situation, and is doing her best to remedy it by quota laws and the like.

It therefore behoves us to take notice ourselves. Australia and Canada will certainly refuse to be the dumping-ground of refuse from England or anywhere else, and it is to be hoped that South Africa will be equally firm. We must look to it that, now the western dumping-ground is gone, our own hospitality does not lead to a still further lowering of the mental standard of England, already dangerously lowered through the follies of ill-judged legislation, by the acceptance of the dregs of humanity that America is rightly rejecting.

When Malthus committed himself to a mathematical formula, that while the population tended to increase in a geometrical progression the food-supply increased in an arithmetical progression only, he made an unfortunate mistake. It is not impossible that his contention is true over a sufficiently long period; or at any rate that it was true before the possibility of synthetic foodstuffs arose; but over the short inventive period of the nineteenth century it was notoriously falsified; and this led to a disastrous rejection of his whole doctrine.

Nevertheless it has always been true, and probably will always remain true, that, in the long run, population tends to overtake food-supply. In the nineteenth century the food-supply actually increased more rapidly than population; but this cannot occur again.[1] It was due to the accident of increased facilities of transport which led to the filling of vast empty lands. Those lands are now nearly full. Canada, America, and the Argentine could produce enormous increases of food; Russia was producing more than she needed and was increasing her cornfields.

The optimists point out that there is still Africa and some of Australia; and anyhow chemical science will go ahead and supply all that is necessary. It is doubtful. At present the food-supply *per capita* in the United States and Canada, the greatest food-producing countries, is actually decreasing, not increasing.[2] In another hundred years or so, at the present rate of increase, America will be full. Allowing for Australia, South America, and Africa—very far from full, but largely unsuited to white races, and with great desert tracts whose irrigation is improbable—the most optimistic estimate can allow three centuries at most for complete saturation—if anything like the present rate of increase is maintained. Indeed in the case of Australia it is vain to hope that a

[1] If the natural rate of increase of population is maintained, of course.

[2] The present glut of wheat is due more to a failure in the machinery of distribution than to any real excess. Economic conditions have limited consumption, but there are large areas which are suffering from hunger while the wheat they cannot obtain lies unsaleable.

population much larger than the present one can be supported within the next generation.

Before the development of transport, nations always tended to reach a saturation-point with a stationary population. That population was thin or thick, according to the general conditions of living. China has for centuries been full. This does not mean that there is no elimination of the unfit—on the contrary, the struggle for existence in China is very severe, and the death-rate very high.

The population problem of Japan at the present day is well known. The efforts to cope with the expansion of Europe towards the end of the last century by the colonisation of Africa are to be found in every history of the period; and the fact that Bismarck encouraged such efforts with the primary motive of giving time for the consolidation of Germany does not alter the biological pressure that lay behind economic and political movements. The inadequacy of the safety valve, especially for Germany herself, was probably not foreseen by the Iron Chancellor; but it was a fundamental cause of the Great War. The place in the sun was no figure of speech, but the need of an over-expanding population.

The sudden gain of control over the environment, giving the whole face of the globe to play with instead of a few thousand square miles; an entire revolution in the methods of production by inventions which alter and speed up industry; these may veil biological factors for a few years, but they cannot alter them.

The world is approaching saturation. We do not want the old correctives, famine, pestilence and war. The first two were eugenic, and would continue to be so, but future wars look like being even worse from the eugenic point of view than the last; and that was bad enough. A war where the cities will be no safer than the trenches, and where even the females of the good stocks are not exempt from poison gas, would be a very different thing, biologically, from the old hand-to-hand encounters, where the strongest man killed his opponent. One begins to sympathise with the Pestering Popinjay in *Henry the Fourth*, who would have himself been a soldier but for this vile gunpowder! But anyway, civilisation cannot contemplate the regulation of population by the old methods, with new horrors added.

There have been two events in the biological history of the world that were of prime importance. The first was when the vertebrates began to adapt themselves to breathing air; the second was when, in the last half-century, man abolished natural selection in the human species.

If we want to save democracy, and turn it into something more rational and better than it is at present, we must see that there is a democracy worth saving.

All of which reflections go to show that we must limit the population—which we are already doing; and that we must limit it by eliminating the worst, instead of the best as we are doing at present. No doubt the more primitive races will gradually vanish, except in places where they are better fitted than others to live. The Red Indian and the Australian Black-fellow are disappearing, and others of the less well-adapted races will follow. The birth-rate of the American negro has begun to decline already.

We must also realise in time that, even if we have an A1 population, saturation means a tremendous international explosion. Rising tension between two nations will continue to have in the background the biological fact that there is no room in the world for all the members of both those nations. Nature takes a very firm line about saturation, even unaided by man's jealousies—a line which makes the world singularly unsafe for such a democracy as we have lately aimed at. But we must not neglect the possible danger of over-limitation. Another century or two might find the Western nations with populations so reduced that, despite their inventiveness, they could not oppose an efficient barrier to some denser population from the east or south; while a single nation which over-reduced its population would be in grave danger from others. Quality is of the first importance, but quantity is not unimportant. If we let our level run down too far, whether in quantity or quality, it is always possible that the remedy will be found in the sweep of a vigorous, less civilised race over the civilised world. History has a way of repeating itself; and civilisations are more fragile than we think.

Any race which was sufficiently ruthless with its own people, sufficiently numerous, and sufficiently aware of the possibilities, could in a few generations produce conditions that would lead to a great migrational explosion which softer and more comfortable nations might find it hard to check, even allowing for the modern advantages of communication and mechanical invention. Improbable though this may sound, it is not beyond the bounds of reasonable possibility.

Chapter VIII

HEREDITY AND INTERNATIONAL PROBLEMS

We may now turn to the inbreeding and outbreeding of nations. The British Isles give an excellent starting-point. We may begin with some statements which, if rather sweeping, are none the less generally accepted by experts as roughly true. There can be little doubt that the most efficient and adaptable race in the world populates the north and east of England and Scotland, though the French are not far behind. The next best is the North German, practically as good as the French, but intellectually rather less adaptable and quick. (We omit America as too recent to have developed a definite character of her own.) On the other hand, leaving out Ulster and a stretch along the east coast, Ireland is inhabited by one of the least efficient populations in Europe; yet a population of many virtues which, when admixed with English or Scotch blood, gives as fine a type as could be desired. But the pure Irish stock has produced hardly a single person of real eminence. Why this strange contrast?

At the end of the Palaeolithic epoch a comparatively low Mongoloid stock, the Iberians, covered Europe. Another wave of a better Mongoloid stock followed these—the Turanian. Then followed a great drive of the Aryan peoples, which reached the West perhaps about 2000 B.C. Following them came a second wave of Aryans, and with them the Celts, of rather low grade; followed a few hundred years later by a second wave of Celts. This last wave of Celts swept over the British

Isles, including most of Ireland. They were rather an aristocratic race, though perhaps less efficient on the practical side than on the intellectual and artistic; and they did not interbreed very much with the lower races established before them. They killed off most of the Iberian and Turanian stock in England and in Ireland, except in the difficult fastnesses of the west.

Then came a finer race of people, fairly closely allied to the last, and still of Aryan stock; tribe after tribe which fought, intermarried, and recombined; kaleidoscopic tribes which swept over northern and western Europe: Germanic, Slav, Frankish, Gaul, Angle, Saxon. All were Aryans; and the northern portions, constituting the Nordic stock (a race very much crossed, but crossed with peoples nearly allied to itself), swept over England, from the fifth to the eighth centuries mainly, and very largely wiped out the Celtic stocks; partly by slaughter; partly, especially in the west and north where the Celtic stocks survived longest, by interbreeding. The Roman stock had already come over; and although it is always said that the Romans did not mingle with the English, and though many of the later legions were Gaulish, when it is remembered that for several hundred years legion succeeded legion in England, and stayed long, the conclusion that there must have been a good deal of interbreeding is almost inescapable.

Whether this is so or not, the Roman stock became merged; so that when the Norman Conquest took place it did not mean the bringing in of a very different race, but rather of another closely allied Nordic stock.

In Ireland the story is quite different. The Danes tried to take hold, and there was a succession of Danish kings who carried on a perpetual and rather unsuccessful warfare from the eastern border. In the centre the old aristocracy fought and ruled. The Iberian and Turanian remained in the west; to some extent free, to some extent as serfs. Then came the English invasion, and one of the blackest chapters of English history—a story of tyranny and murder which killed off most of the Celts; and the chapter closes with the unforgettable cruelties of Cromwell, which finally took the land from the descendants of the Celtic rulers. Eastern Ireland became largely English in blood, at least among the upper classes; though the older race did to some extent re-establish itself in a mongrel form. Politically the east became ultra-Irish. But all along the western coast remains to this day a fairly pure Ibero-Turanian race, with perhaps just a dash of Dane, Spaniard or English here and there. That stock has been inbreeding since prehistoric times—a nearly pure stock of an inferior type. Nothing great comes out of it.

Set over against this inbred stock the English and Scotch and note the contrast. These have arisen through the continuous intercrossing of allied stocks, followed by a period when communication is not easy, so that inbreeding takes place in the towns and villages. First outbreeding of good stocks, then inbreeding with Mendelian segregation. In ten generations you begin to get an extraordinarily virile and enterprising race, for the unsuccessful variants have been eliminated in the struggle. The English race begins to come to the fore.

All this is highly suggestive. Close inbreeding of a primitive stock is not a success; a free outbreeding of allied stocks, followed by close inbreeding, gives heterosis vigour.[1] This clue is worth following, for it has bearings of national and international importance especially for a country which has many races of different colours to rule.

Following it up, we must not however in our new-found enthusiasm for purely biological explanations forget the importance of the cultural or epigenetic factors. To neglect those would be as great, though not as disastrous, a blunder as to neglect the importance of the germ-plasm.

People say, Look at the Jews! What a fine, inbred stock it is! Nothing could be farther from the truth, biologically speaking. The Jew is not a pure-bred stock. He started probably as a cross between the Hittite, the Amorite, and the Semite. He got his nose from the Hittite; his culture to a great extent from the Arab; his vigour, perhaps, from the Amorite. After this mixture he inbred for a long time, manifesting the heterosis vigour that might be expected from his ancestry. Then, in the Hellenistic period, he began to proselytise with extreme vigour and success. The Jew of the Diaspora was a cultural, not an ancestral Jew: if you are a Jew by religion you can intermarry with Jews. Biologically this makes all the difference. The Spanish Jew is as different from the Polish Jew as the Spaniard is from the Pole—or nearly; and the Hittite nose is frequently absent. Intellectually, the German Jewish stock is perhaps the finest. There is in it a good deal of the true Jewish blood, but what races contributed I do not know.

History, and especially early history, is the record of the migration of races and their intermingling with other

[1] We may again refer to the suggestive material given in Petrie's *Revolutions of Civilisation*.

races. Their subsequent ups and downs have more to do with the germ-plasm, either of the ruling classes, or of the nation as a whole, than is generally realised. The problem now facing the world, thanks to freedom of transport, is the question of the intermingling of the big divisions of the human race. We have already seen that the crossing of closely allied races is of great benefit, if followed by a period of inbreeding. Is this equally true of races farther apart?

In the early days of man there was fairly complete segregation; and various attempts at human beings came into existence. Climatic conditions—first the pendulum swing of the Glacial epochs, and later the northward shift of the rain-belt in the fifth millennium B.C., led to migration on a scale larger than that imposed by the search for grazing-grounds, which had become an ingrained habit imposed by the domestication of cattle. History is the story of these migrations, and of the conquests and interminglings of races. But until recently there were still geographical barriers as well as social and psychological ones; and three chief types of *Homo sapiens* became differentiated—the black, the yellow and the white—though only this one species of man seems to have survived out of several experimental types. Now that the geographical barriers are nearly negligible, it behoves us to decide whether it is advisable to mix these main varieties. Too much stress must not be laid on the unsatisfactoriness of many past admixtures; for as a rule they were due to the mating of rather undesirable types. The Eurasian began from a poor type of Portuguese and a poor type of Indian—Goanese, Tamil or Hindu; and the ancestry of later Eurasians is hardly better. The half-caste of South Africa came largely from the "poor white". But the half-caste Indian of Central and South America often had good Spanish blood; and here too the mixture is not a success. The Dutch-Malay blend seems fairly sound, but it may be doubted whether it is as good as the pure Dutch.

The problem of the black race is the most urgent; and as it has been fairly well studied in America, we have sufficient data to base our discussion mainly on that.[1]

The black race is inferior, mentally and physically. This is not a matter of colour-prejudice: the intelligence tests for the American army during the Great War showed it very clearly. The men were divided into seven grades. Of 94,000 whites the percentages of the successive grades, from the highest downwards, were 4·1, 8·0, 15·0, 25·0, 23·8, 17·8, 7·0. For the blacks the percentages in the grades on the same tests were 0·1, 0·6, 2·0, 5·7, 12·9, 29·7, 49·0. There is no question at all that these figures, whatever allowance be made for prejudice and unsatisfactory tests, do indicate a much lower mental grade in the negro. The study of more or less eminent negroes by Reuter shows that practically every individual in the list has some white blood, and there are ten times as many with at least 50 per cent. of white blood, as with 25 per cent. or less. The negro *Who's Who* published in 1916 has at least 97 per cent. of mulattoes in it. Indeed Paul Dunbar, a poet little known on this side of the Atlantic, seems to be about the only pure black of any public fame. The fact of the matter is that there is only one person with any black blood at all who can be put among the world's really great men, Alexandre Dumas, who had a negro grandmother. It

[1] A summary of the chief data will be found in East's *Heredity and Human Affairs*.

may be urged that this is all due to lack of opportunity; but the black Republic of Haiti gave an opportunity. It rapidly became one of the most corrupt places in the world: religion went back to voo-doo, human sacrifice was common, there was no sort of justice, and there was an army in which I believe I am right in saying there were more generals than privates. The Republic of Liberia has also earned an unenviable reputation leading to the intervention of the League of Nations.

The mulatto is very obviously superior to the negro, but quite as obviously inferior to the white.

The plain question is, Is it worth while to raise the negro at the expense of the white? and the plain answer is that, as the blend is lower than the unmixed white, it certainly is not.

Physically the negro is fairly successful in temperate climates, though very liable to tuberculosis; but the pure negro is beginning to die out in America none the less. His birth-rate is decreasing, while that of the mulatto is not, to anything like the same degree; and his death-rate is not decreasing as fast as that of the white or the mulatto. In measurable time the pure negro stock will cease to exist in America.

But what of the yellow races? In China and Japan we have virile races, of a high intellectual grade, though perhaps not as high as the Caucasian. The problem is whether you are going to encourage or even allow cross-breeding with the white.

The great objection to crossing with the negro stock does not apply here, for there is no such conspicuous inferiority. But the natural repulsion to the idea must not be forgotten, for it is very widespread and *may* be the psychological manifestation of a biological incompatibility.

The yellow races have inbred for a long time, after various Mongoloid blends had taken place, and have suffered an extremely severe struggle for existence. Mendelian segregation must have gone a long way; and the factors carried by the forty-eight chromosomes must be well matched. This must also be true of the factors carried in the nuclei of the black and the white races respectively. But it by no means follows that if you take twenty-four chromosomes from the yellow race and blend them with twenty-four chromosomes from the white race a well-adapted hybrid will result. It is extremely unlikely. Great heterosis vigour might occasionally occur; but the mixture would almost certainly produce a large number of inferior and ill-adapted types. Possibly, if we could face a thousand years of adaptation by severe competition which eliminated the biologically unfit, we might in the end get a hybrid race which preserved and even intensified the best qualities of the yellow and the white, and from which the weaker factors were eliminated; but one cannot look upon that thousand years with equanimity.

It all boils down to a simple question of inbreeding and outbreeding, and the old story of Miss King's rats. If you inbreed a race, and then cross it with another inbred race, the two races being very closely alike, you get heterosis vigour, and a fine stock. This is what happened in the admixture of nearly related tribes which gave rise to the Anglo-Saxon race.

But if you cross two very different races you may upset the harmony of the genes, and then your selective adaptation has all to be done again. You may get great heterosis vigour—or you may not!—but it is liable to be rendered useless by some incompatibility of factors. It will be remembered that our radish-cabbage was sterile, for instance, while our homozygous yellow mouse could not live at all.

For a nation to outbreed with a nearly allied nation, and then to inbreed, has proved a vast advantage, for the general balance of the genes is not upset; but to outbreed with another fairly homogeneous race whose genetic constitution is markedly different from your own will probably upset the balance badly; and unless you are prepared for a long and strenuous process of selection it had better be left alone. It is by no means certain that, after all the selective struggle, the result would really be an improvement!

No nation can be built without its element of misery: the recessive genes exist, and the unhappy legion is bound to emerge; but we are morally responsible for seeing that the unhappy are kept down to a minimum; and this is best done by avoiding both the breeding of the unfit and the introduction of disharmonies which will upset the general genetic balance.

Considering civilisation as a whole, and deciding that its influence depends upon the action of epigenetic factors upon a germ-plasm which may be fit to use them, or may be unfit, it becomes clear that what is of the highest importance is to preserve the usefulness, and indeed the very existence, of those epigenetic factors in the only way possible: that is by keeping up the supply of sound germ-plasm, and, as knowledge gives increasing chance and responsibility, by improving that germ-plasm through a rational selection which can less wastefully fill the place of natural selection—selection before birth instead of after.

Education is really only a sieve, not a means of im-

proving the quality of a people. Art, literature, philosophy, science, invention, government, and finance must
always be the work of the few who are best endowed, by
constitution, not by birth; but their appreciation can
and ought to come from the many. If we spend all our
time giving the wrong sort of education to people who
are not capable of profiting by that particular sort of
education, turning out an inferior black-coated class
who despise the manual activities they have not learned
to understand, yet have no chance of rising in their
unsuitable *milieu*, we are only wasting time and money
and material. Some should be educated for agriculture,
some for trades, some for domestic work, some for
clerical work, some for the professions; and only those
fitted should be trained for each. But all should be
trained for citizenship and appreciation. It is no question of merit: it is merely a matter of aptitudes. At
present the selective machine of education is badly
adapted, sterilising those it does select, and hampering
the development of all by treating all in much the same
way.

The old idea of a social ladder will have to be replaced
by the idea of social bridges. One of the tragedies of
democratic legislation has been the conversion of a caste
system into a biological entity. If we eliminate the unfit,
slowly and steadily, it will be possible to develop the idea
that no occupation is more honourable than another.

The menace of perpetual motion as an end in itself—
whether the perpetual motion of a ball in a field, of a
dancer in a ball-room, or of a man and a maid in a car—
will fall into place when children are valued at their
proper rate, and when the wicked expenditure on the
useless is replaced by an adequate training of a lesser

number of really valuable citizens. The idea of the policy of the nation being determined by those least fitted to understand it will become part of the comic history of England, when adequacy in your own line of life is the qualification for one or more votes.

Legislation which disfranchised the individual while that individual cost the state more than he contributed would almost certainly be biologically sound. One individual one vote presupposes the doctrine that all men are equal, and then obscures the issue with an ethical idea which is really irrelevant. *Sub specie aeternitatis* men may be of equal value; for the purpose of guiding the State they most certainly are not.

A sound germ-plasm will soon create a sound education and sound politics; but the importance of education will become not less but more, for no character can develop without the appropriate stimulus. Nothing can lessen the vital importance of good conditions if there is to be a good population.

Prisons, asylums, special schools, workhouses, reformatories, refuges, will not cease to exist, but they will be few, and will grow steadily fewer. Their cost will be nothing in comparison with what it is to-day. Unemployment will be far less, and governed only by the ordinary trade cycles—and even these we may have begun to understand and to control.

A Utopian vision? Perhaps; but not in the least impossible of realisation, if we will only learn before it is too late that only upon a sound inheritance can you build a sound nation. The lessons of history are clear to read, and the biological knowledge necessary to explain them is already in a large measure ours. But the sands are running out.

We seem to have got away from international matters again; but in fact we have not done so. Closeness of contact between nation and nation, brought about by facility of communication and transport, makes it increasingly necessary that every nation shall be contented and sound. An inefficient race is a perpetual temptation to the instinct of expansion. If wars are to be avoided, over-pressure of population must not exist in any nation; and this is only possible for a vigorous race if the national stock is sound throughout. If there are many economically unfit to be supported there is need for many to support them; wages must go down, or be uneconomic; and both new markets and new sources of raw materials must be sought. Economic competition is intensified; and the line between a pressing economic competition and envy, hatred, and malice is thin.

There are two ways of producing uneconomic populations; one is by inbreeding inferior stocks through misdirected selection; the other is by producing inferior stocks by unwise outbreeding.

The publication of Professor Julian Huxley's important book, *Africa View*, makes a further note on the black races desirable, since that book is likely to affect, and perhaps to determine, our African policy. It must be conceded that the numerical data suggesting the mental inferiority of the African races are American, and are thus both coloured by race-prejudice and limited to the groups which supplied slaves for the plantations. But when all allowances are made this evidence does show that by western tests the African is definitely inferior to the white. The *relative* numbers in the different grades are conclusive, however much the scales were unconsciously weighted. The African may be well adapted to his own life and culture in tropical Africa, but changed that life and culture is bound to be. The main point of Professor Huxley's book (with which one must heartily agree) is that only by educating him along the lines natural to him can we give him his best chance; yet this does not alter the fact that by white standards he is found wanting; indeed it tacitly concedes the point.

Nevertheless he is often an excellent agriculturalist, and along his own lines he may go far. It is our business to give him his opportunity—as in many places we have been doing for some time. Of course, the argument against cross-breeding, owing to the disharmony of genes likely to result from intercrossing, remains unaffected by any possible existence of prejudice in the compilation of intelligence-test statistics.

Chapter IX

CONCLUSION

It is not the business of science either to point a moral or adorn its tale. Yet, since I have already outrun discretion by some moralising, the matter will be made no worse if I add some brief reflections by way of finish and adornment. The history of Europe in the last hundred and fifty years, which have after all seen greater changes than the whole six hundred thousand to a million years of man's previous existence, shows clearly that progress lies along the lines of Democracy. In France the political revolution, in England the industrial revolution, were the birth-pangs of man's greater freedom. 1814 saw, at Vienna, the vain attempt of Western civilisation to put back the clock. Till 1848 the infancy of freedom was on the surface a struggle with the diseases of society whose issue was gravely doubtful; few realised the vitality of the new order.

From 1848 to 1918 Europe saw the true contest between the old and the new ideas. The central empires clung to despotism, while the rest moved slowly and painfully towards freedom, making costly blunders whose results are with us to-day. There could be no doubt of the issue. The tragedy of 1914–1918 was inevitable, and happily Democracy won outright, and the world was spared that second effort which failure would have entailed. But Democracy can impose its own tyranny. If it does, and if I read aright the history of evolutionary progress as the effort of the organism to make a fuller use of its environment; to gain more control; to be more free; a worse struggle is yet to come. In that struggle civilisation may be destroyed. Whether a new and better civilisation will arise is uncertain; probably it will. But centuries of misery must intervene; and no one can contemplate the prospect without a pang of dread. Many organisms in the past have failed and died; evolution is a blind process. Yet in the whole economy of living things we do see progress; if not in this race then in that; and the progress is progress in freedom. In life there is an upward movement.

What Democracy has won, it has won because it aimed at freedom. Nevertheless, the foundation of freedom is the rigid necessity of physical and biological laws. Without such a foundation there could be no true, ordered freedom, but only chaos.

To believe that we can escape the laws which govern the transmission of hereditary characters is as futile as to believe that we can achieve perpetual motion, or stay the stars in their courses.

Democracy has yet to face this truth, and to learn that the unborn are as much one's neighbours as the born. Though a scientific discussion of evolution and heredity is not concerned with the first and great commandment, it is inescapably concerned with our duty towards our neighbour.

If we breed from the unfit and ordain that the fit shall be barren the result is as certain as it is certain that water will run downhill. If Democracy is to win its freedom and achieve its wonderful possibilities, it must both encourage the breeding of the fit and render the breeding of the unfit impossible; its legislators must have vision enough to look into the far future, know-ledge enough to avoid the wholesale destruction of its living capital, and courage enough to resist the shortsighted demands of the ignorant and selfish. An aristocracy of ability, whether of physique or hand or brain, must replace both the aristocracy of position and the dictatorship of votes.

Already it is an easy matter, surgically speaking, to sterilise those whose unhappy inheritance renders their reproduction undesirable. [Ignorance and prejudice are still so widespread in this matter that I may perhaps be pardoned for stating clearly that ligature of the *vas deferens* in the male or of the *Fallopian tubes* (oviducts) in the female has no ill effects whatever, and does not in any degree affect married life except in so far as it renders conception impossible. Sterilisation is not castration, and has none of the tragic results of that operation.] It is often argued that, since we have as yet no means of recognising the presence of a recessive character in the haploid or latent condition, the effects of sterilising the patently defective would be small. As a matter of fact it would certainly mean a by no means negligible reduction; probably from 11 to 20 per cent. in a single generation. But since defective tends to mate with defective or low-grade, and low-grade with low-grade, it is probable that this sexual selection would result in a markedly greater reduction of the "problem-class" than is anticipated, while sterilisation of the exhibitors of mental defect would also lessen the carriers in the next generation[1]. But sterilisation cannot stand alone; it must be accompanied by a training that will

[1] Since we are applying selection to eliminate the unfit, Hardy's law, which shows the paradoxical fact that when you have a dominant and recessive breeding at random, the proportions of the two homozygous and the heterozygous types will be stable at several different values, does not affect our argument. The matter is too abstruse for discussion: an outline of it will be found in the new edition of Punnett's *Mendelism*.

render the less seriously defective capable of useful and happy, if slightly limited, lives as citizens under friendly guidance. For the hopelessly defective there must be institutional care. But as time goes on the numbers of the unfortunate will decrease.

But hand-in-hand with this negative process of diminishing the incidence of unfitness must go the positive process of encouraging the breeding of the fit.

People do not yet realise that between 1940 and 1950 the population of the British Isles will be stationary, and thereafter will decrease at a rapid rate. Some decrease is probably desirable; undoubtedly a check on breeding is necessary to counterbalance our modern methods of saving and prolonging life, if wars and pestilence and perhaps famine are not to play their old terrible rôle in natural selection. But to my thinking the signs point towards not only a qualitatively disastrous inferiority, but also a dangerous shortage in the quantity of population in western Europe before another hundred years are passed. Such things have happened before, time and again; and the result has always been the migration westwards of a less civilised race, centuries of darkness and strife, and then the gradual emergence of a new civilisation. Exactly what this would mean under modern conditions no one can forecast; but I at least cannot think of the possibilities without dismay.

If our present civilisation is to survive and develop it is undoubtedly necessary that we should preserve a population of adequate density and of high quality; and if we are to do this we must set about encouraging the breeding of the fit without delay. Already we have reached, or passed, the danger-point in the matter of quality; and it looks as if it might be only a few decades before we reach it in the matter of quantity also.

Measures are not hard to devise. The essential is that we should encourage reasonably large families among those of our citizens whose biological worth is proved by their social efficiency. We have only to look to the system so successfully operating to-day among the "ouvriers" of France, which ensures, by a self-supporting pool within the industry, that family allowances proportional to the earnings of the father shall prevent the lowering of the family standard of living by the birth of several children, to gain a hint of one possible method. But we are straying into the realm of Political Economy, and it is time to stop.

Within another half-century Democracy in England will see the fingers of a hand writing upon the wall, unless we abandon our callous feasting and apply ourselves to the responsibilities of government with knowledge. Once the words MENE, TEKEL, PERES are written so that all can read, it will be too late. I do not believe that day will come. Behind the blind mistakes of Democracy is a force that expresses consciously that struggle of the creature towards freedom which is the one evidence of progress in evolution. It is not too late yet.

But we may not forget that only one stem followed the zig-zag upward course; the branches of the tree of evolution may be useful and well-adapted, but they are composed of a million organisms which failed to progress beyond a certain limited stage,—failed to win more than a very little control over their environment. Some are extinct, some still survive, but none are progressing. The human race is moving upwards; our new knowledge is evidence of that movement; but the progress of man is conscious, and consciousness means responsibility.

GLOSSARY

allelomorph: a factor inherited alternatively to another factor; e.g. tallness and shortness in peas. Though apparently there is a pair of alternative characters, the effect is due to the presence or absence of one factor.

amnion: a membrane enclosing the embryo in the higher vertebrates.

analogous organs: organs in different animals which have the same function, but are not based upon identical structures. Similar in function, they differ in origin.

annelida: the phylum which includes the marine worms and earthworms.

archenteron: the primitive digestive cavity of the embryo.

arthropoda: the phylum which includes the Crustacea, Insects and Spiders.

asymptotic curve: a curve which becomes indefinitely nearly parallel to one or both of the coordinate axes, but only achieves parallelism when produced to infinity.

atavism: reversion to an ancestral type.

blastopore: a small opening which leaves the archenteron in communication with the exterior when the *gastrula* is fully developed.

blastula: a hollow sphere, sometimes ciliated, which frequently results from the segmentation of a fertilised ovum; it constitutes a definite stage in the development of a larva or embryo.

catalytic agent: a substance which speeds up a chemical reaction, while generally remaining itself unchanged at the end of the reaction.

cell: a vague but convenient term denoting the structures which form the units of a tissue. A cell consists of protoplasm surrounded by walls frequently pierced with fine holes, having in it a complex structure of denser protoplasm called the nucleus. A cell may contain many non-living substances, such as fats, sugars, ferments, starch, either solid or held in solution; and sometimes large vacuoles filled with liquid. Cells differ greatly in their degree of independence.

chemotropism: a movement in response to a chemical stimulus (see *tropism*).

chlorophyll: a group of four pigments which give a green colour to most plants. They absorb light of certain wave-lengths, and the energy of these waves is used by the plant in building up CO_2 and water into sugars, etc. (photosynthesis).

chromatin: the darkly-staining material in the chromosomes. The term is convenient, but without an exact significance. It is purely descriptive. Though it is possible that the dark staining is due to a definite chemical substance, it is certain that if this is so the substance can only constitute a small fraction of each chromomere, which must consist of many different substances; perhaps, however, there are many substances in the chromomeres which have an affinity for the dye haematoxylin.

chromomeres: the granules of chromatin which form the chief constituents of each *chromosome.* (See *nuclear division.*)

chromosomes: small darkly-staining rod-like fragments of nuclear material which appear when a cell is about to divide. They are definite in number, occur normally in pairs, and bear the hereditary factors. (See *nuclear division.*)

cilia: mobile, hair-like protoplasmic projections of a cell, which act as oars or set up currents.

coelenterata: the phylum which includes the sea-anemones and jelly-fish.

coelom: a hollow organ originating as a pouch or split in the mesoblast. It usually, but not always, forms the main body-cavity in which the gut lies; and its walls are intimately associated with the formation of the genital and excretory organs.

colloid: a chemical substance which when in solution will not diffuse through a parchment membrane. The molecules are large, and are grouped together as micellae, so that they cannot pass through the pores of the membrane. Colloid solutions are intermediate between ordinary solutions of crystalloids and suspensions such as gamboge in water. Colloids can exist in two phases, as *sols* in which the solid particles are floating in a fluid solvent, and as *gels* in which droplets of the solvent are inclosed in a jelly-like matrix formed by the micellae.

conditioned reflex: a chain of *reflex actions* (q.v.) in which an intermediate stage is suppressed, e.g. if a child is offered something with an unpleasant taste, the first time the object will be placed in the mouth, and then spat out with disgust; the second time the object will be refused with an action similar to spitting. Pavlov attempts to explain all human activities on this basis.

conjugation: a process by which in the Protozoa and Algae a portion of the nucleus of each individual unites with a portion of the nucleus of the other, after some nuclear material has been discarded. This process is the forerunner of true sexual reproduction, in which the nucleus of the spermatozoon fuses with that of the ovum. (See *gamete, zygote, nuclear division.*)

cytology: the study of cell-structure.

cytoplasm: the general protoplasm of a cell which surrounds the nucleus.

diploid: in the diploid condition the chromosomes are in pairs; in the formation of the germ-cells the pairs are separated by the *reduction division,* so that each gamete has only half the normal number and is said to be *haploid.* We here neglect the X-chromosome. (See *nuclear division.*)

embryo: a developing organism which is still dependent on the yolk of the egg or on the blood of the mother for its food-supply.

epiblast: the outermost layer of cells in the embryo, giving rise to skin, central nervous system, etc.

epigenetic factors are those which affect the development and life of the organism but play no direct part in heredity.

epithelium: a layer-tissue one or more cells in thickness covering a surface or lining a cavity.

exogamy: mating outside the family or clan.

exogenous: produced by outside causes.

formaldehyde: formalin H—C—H. Probably an intermediate stage
$$\overset{\|}{\text{O}}$$
in the photosynthesis of sugar from CO_2 and water in green plants.

gamete: a sex-cell. The male gamete is known as a *spermatozoon,* the female gamete as an *ovum.*

gastrula: the stage of development which succeeds the blastula. One side of the blastula is pushed in, or *invaginates,* to form a two-layered cup. The outer layer is the *epiblast;* the inner, which lines the archenteron, is the hypoblast.

gel: see *colloid.*

gene: a convenient if linguistically indefensible term for a Mendelian factor.

germ-plasm: Weismann introduced the terms *germ-plasm* and *somato-plasm* to distinguish the protoplasm of the germ-nucleus from the protoplasm of the body-cells. The terms are convenient, though strictly they depend upon a definite theory of inheritance which is not generally accepted without some reservations.

gonad: the reproductive organ. The male gonads are called *testes*, the female gonads *ovaries*.

haploid: see *diploid.*

heterosis: practically amounts to cross-breeding. The word is used chiefly in the phrase *heterosis-vigour*. This expresses a frequent result of cross-breeding between races not closely related, and conveys the idea of vigour caused by the presence of many desirable Mendelian factors in single doses, instead of a smaller number in double doses. (See *homozygote* and *heterozygote*.)

heterozygote: when both members of a pair of allelomorphs are present in the same organism; or when, in the phraseology now used, a Mendelian character is present in a single dose, the organism is called a *heterozygote*, and is described as *heterozygous*.

homologous organs: organs in different animals which may or may not have the same function, but whose origin and fundamental nature are the same.

homozygote: when the Mendelian character is present in a double dose the organism is *homozygous*.

hormone: a chemical body of fairly simple structure secreted by certain glands or cells, which, poured into the blood-stream, has a specific influence on some other organ or tissue. The most familiar hormones are those secreted by the ductless glands such as the thyroid and pituitary, and the insulin secreted by a tissue embedded in the pancreas.

hypoblast: the innermost germ-layer which lines the alimentary canal.

karyokinesis: see *nuclear division.*

larva: a developing organism which obtains its own food, though it has not yet reached the adult form.

meiosis: the process by which the number of chromosomes in the gametes is reduced to half that characteristic of the parent organism.

meiosis and *mitosis:* see *nuclear division.*

mesoblast: the germ-layer intermediate between the epiblast and hypoblast, from which muscles, blood-system, coelom and urinogenital organs arise, as well as the skeleton in vertebrates. Not present in the lowest metazoa.

metazoa: animals whose bodies are divided into cells and tissues.

micellae: the groups of molecules in a colloid sol.

monotremata: the most primitive surviving order of mammals. They lay eggs, and have several reptilian characteristics. *Ornithorhynchus* and *Echidna* are the only existing representatives.

mutation: a sport or considerable variety which appears suddenly owing to a change in the genetic constitution: the characteristics are inherited.

notochord: a stiff gelatinous rod lying beneath the nerve-cord which acted as a stiffening skeleton in the ancestors of the Vertebrates. It is later replaced by the spine or vertebral column, but always appears in the embryo.

nuclear division: see end of Glossary.

ovary: the female gonad or sexual organ in which the ova are formed.

ovum: the female gamete or sex-cell. It is comparatively large, and usually contains besides the nucleus a store of food material, the *yolk*.

peristalsis: the muscular activity of the intestine by which the food is passed down. Rhythmical constrictions preceded by relaxations travel slowly along the intestine, forcing the contents backwards.

pharynx: the region of the alimentary canal just behind the mouth: the cavity of the throat.

phylum: one of the main groups of the animal kingdom. About a dozen phyla exist.

placenta: the organ by which the embryo of a mammal is attached to the uterus of the mother. It is highly vascular, and food and oxygen pass from the blood of the mother to that of the embryo.

protoplasm: a general term for the living matter of which plants and animals are composed. It is not a chemical substance, but is a colloid aggregate of proteids and other organic compounds in a peculiar physical state.

protozoa: the lowest group of animals, often miscalled a phylum. The organisms composing it are mostly microscopic; their bodies are not subdivided into cells and tissues, though there may be a considerable amount of organisation and physiological division of labour in the protoplasm of which they are composed. *Amoeba* and *paramoecium* are familiar examples.

reduction-division: see *nuclear division*.

reflex: when an impulse set up by some stimulus travels from the sense-ending of a sensory nerve to a cell (*neurone*) in the spinal cord, leaps across a tiny gap to the processes of a "motor" cell, and causes an impulse to travel out along the motor nerve to a muscle, causing this to contract (no reference to the brain, or to consciousness, being primarily involved), the action of the muscle is called a *reflex action*, and the whole circuit is called a *reflex arc*. A familiar example is the knee-jerk caused by crossing the legs and sharply tapping the upper one just below the knee-cap. Reflex actions will take place even in a vertebrate whose brain has been destroyed. The gap between the fine processes of the nerve-cells across which the impulse leaps, constitutes the *synapse*.

somatic: pertaining to the body. Somatic cells are distinguished from *germ-cells*.

somatoplasm: see *germ-plasm*.

spermatozoon: the male gamete or sex-cell, consisting chiefly of a nucleus and a vibratile "tail".

synapse: the gap between the processes of the nerve-cells or neurones.

testis: the male gonad or sexual organ.

tissue: a group of cells associated for the discharge of some definite function; cells mainly of one type as a rule.

tropism: protoplasm responds to stimuli. An organism such as a protozoon may move towards light or away from it, towards a drop of dilute acid or away from it. Such movements, which are not coordinated by a nervous system, but are due to unequal stimulation of different parts of the organism, are called *tropisms*. An organism is said to be negatively phototropic, chemotropic, etc., according as it moves towards or away from light, a chemical stimulus, etc. Loeb attempts to base all animal movement upon tropisms; the reflex arc would be regarded as a later development of less importance.

ungulata: the hoofed mammals.

zygote: the product of the union of two gametes. In metazoa the term is equivalent to *fertilised ovum*.

nuclear division (otherwise called *karyokinesis* or *mitosis*): when a cell is about to divide a network studded with granules of *chromatin* appears at the surface of the nucleus just beneath the nuclear membrane. There may be in the network one or more larger aggregations of chromatin also. A small star with radiating lines, having at its centre a deeply-staining granule known as the *centrosome*, appears in the cytoplasm. The nuclear network now becomes a thread, apparently continuous, by absorption of the cross-meshes of the network. This is the *spireme*. The centrosome divides, and a field develops resembling that between two opposite magnet-poles. This constitutes the *spindle*. The spindle-fibres undoubtedly indicate lines of stress in the cytoplasm, and possibly the force between the poles (centrosomes) (whatever its nature) has a physiological effect, altering the state of the colloids of the cytoplasm so as to form actual threads. (In a cinema-film taken under the highest powers of a microscope, and speeded up, the pull of these lines of stress upon the chromosomes (*vide infra*) is vividly shown; though the existence of a pull does not settle the question of whether actual threads are present.)

In the next stage the spireme-thread thickens and breaks up into a definite number of rods—the *chromosomes*—whose number varies from 2 to 168. This number is perfectly regular in any given organism (we here neglect the question of the X-chromosome). The shape of the chromosomes varies from a thread to a rounded blob. There is conclusive evidence that the chromosomes carry the hereditary factors, and that the rods really exist in pairs, though the fact is not always obvious. The nuclear membrane now breaks down, and the spindle stretches across the nucleus.

The chromosomes often become **V**-shaped, and may arrange themselves in the middle of the spindle with the arms pointing outward like a star: they may merely assemble rather formlessly in the middle (*equatorial plate* stage). The **V**'s begin to split, each granule of chromatin (*chromomere*) dividing into two. Half of each longitudinally-split chromosome is drawn to one pole along the spindle-threads, the other half to the other pole (*vide supra*); at each pole the daughter-nucleus thus formed goes back, more or less through the spireme and network, to a resting stage. The spindle-fibres begin to thicken in the middle and degenerate: the cell-wall becomes pinched in, and after a strange turmoil well seen in the cinema-film the two halves are pulled apart; the centrosomes gradually disappear. Thus we arrive at two daughter-cells each with a nucleus containing not merely a half of each chromosome of the mother-cell, but probably a half of each chromomere or granule. The above description applies in broad outline to the division of every *somatic* cell in plants and animals, except for a few cases where the process is not fully developed or the cells are degenerate. There are minor differences of detail, and various points are omitted in the foregoing account.

In the *maturation-divisions* of the *gametes* the process is more complex, since it is necessary that the number of chromosomes should be halved, as otherwise fertilisation would lead to a doubling of the number in each generation.

[Probably in the original state the organism had, as it were, only half the chromosomes, and when conjugation occurred the number was doubled; for a good many plants which have an alternation of generations show in one—and that the more primitive—generation a *haploid* state, while the more complicated generation, sexually produced, is *diploid*, having pairs of chromosomes, e.g. a fern-plant (diploid) forms haploid spores which give rise to a small *prothallus* which bears the sexual organs. The *zygote* formed after fertilisation becomes the familiar (diploid) fern-plant again. Analogous states are found in the animal kingdom, and cancer-growths show a reversion to the more primitive haploid condition. Thus the *gamete* really represents the original number of the chromosomes.]

In the formation of the gametes there is no radical difference observable until the chromosomes are formed. These conjugate in pairs: exactly what occurs is not fully understood, but the

---

Fig. 46. General scheme of the maturation divisions (meiosis) in the spermatogenesis of *Lepidosiren* From W. E. Agar, *Quart. Journ. Micr. Sci.* vol. LVII, 1911, p. 9; figures kindly lent by Dr Agar. *A.* Resting nucleus, primary spermatocyte. *B.* Leptotene stage. *C.* Threads beginning to pair side by side (Lepto-zygotene stage). *D.* Completion of pairing ("bouquet" stage, zygo-pachynema). *E.* Strepsitene stage, beginning of synizesis. The double threads have split except at their ends, forming long loops. *F.* Late synizesis. *G.* End of synizesis. Bivalent loops are shortening and thickening, and are beginning to break up into their constituent univalents. *H.* The nuclear membrane has disappeared. The loops have broken up into the diploid number of chromosomes each with a transverse constriction. *I.* Appearance of the spindle. Second pairing of the chromosomes beginning. *J.* Second pairing complete; definite bivalents taking their place on the spindle. *K.* Anaphase of first spermatocyte division; bivalents separated into univalents each still having a transverse constriction. Spindles rotating in preparation for the second division. *L.* Metaphase of second spermatocyte division. Univalents, still transversely constricted, splitting longitudinally.

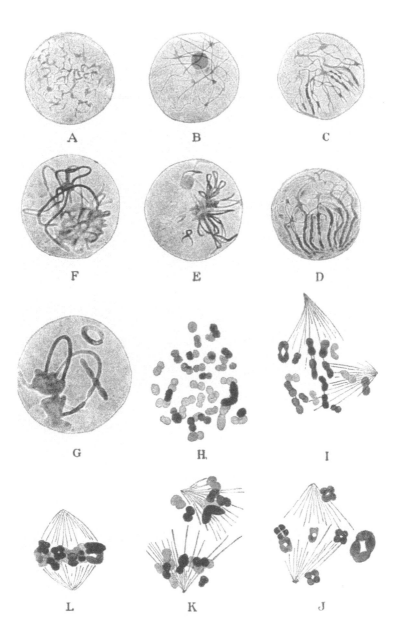

Fig. 46.

fact that they frequently coil round one another spirally is important, as it leads to occasional breakage. Then the chromosomes become more or less V-shaped again, and either form rings by the fusion of the tops of the V's or form groups consisting of four little blobs of chromatin—the *tetrad*. (There is evidence that the tetrad is merely a pair of V's in which the chromatin has all run to the tops.)

Then the *whole* chromosomes draw apart to the poles, so that we get half the number of *complete* chromosomes at the poles, instead of the full number of half-chromosomes; but the chromosomes have previously conjugated in pairs, which may modify them. This stage is known as the *reduction division* or *meiosis*. It results in the formation of two nuclei which are qualitatively different from each other. Without going through a resting-stage another spindle is formed in each cell, the chromosomes split and are drawn apart to the new poles, and the result is four gametes each containing two half-chromosomes instead of four; of these gametes two differ qualitatively from the other two. In the case of the female gametes only one of the four becomes an ovum, the other three being thrown out as two *polar bodies*: this point is, however, unimportant, and is merely a result of the necessity for storing yolk.

In the somatic and germ-cells alike there is almost conclusive evidence that although the chromosomes disappear in the resting-stage and begin to reappear as a mere network of chromatin, yet they retain their identity through innumerable divisions, in a very great degree if not entirely.

Hence the union of ovum and spermatozoon, each having half the usual number of chromosomes, leads to a *zygote* with the normal number; but the separation of whole chromosomes in the meiotic division causes the spermatozoa and ova to differ among themselves; and this makes for variation in the zygote. Probably all the chromosomes carry the factors ordinarily essential to the life of the organism; but each chromosome also carries its own special factors which modify the organism in a greater or less degree.

Figs. 38 and 46 show the stages clearly. Fig. 38 is diagrammatic; fig. 46 closely reproduces the actual stages observed.

INDEX

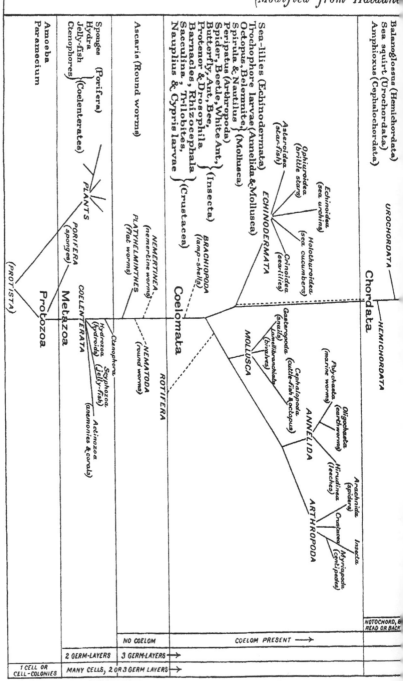

CHART INDICATING PROBABLE F
(Modified from Haldane

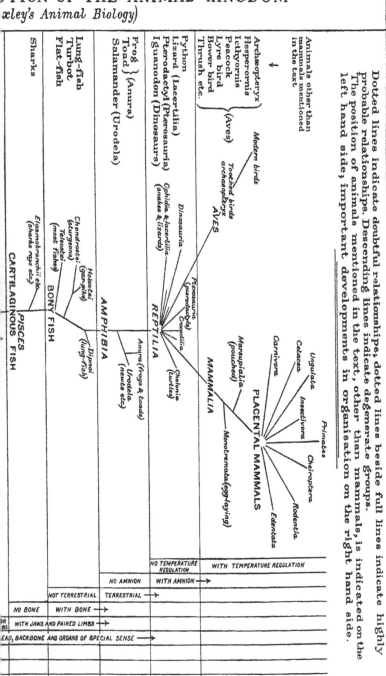

EVOLUTION OF THE ANIMAL KINGDOM

(Huxley's Animal Biology)

Dotted lines indicate doubtful relationships; dotted lines beside full lines indicate highly probable relationships. Descending lines indicate degenerate groups.
The position of animals mentioned in the text, other than mammals, is indicated on the left hand side, important developments in organisation on the right hand side.

Animals other than mammals mentioned in the text

Archaeopteryx
Hesperornis
Ichthyornis
Peacock } (Aves)
Lyre bird
Bower bird
Thrush etc.

Python
Lizard (Lacertilia.)
Pterodactyl (Pterosauria)
Iguanodon (Dinosaurs)

Frog } (Anura)
Toad }
Salamander (Urodela)

Lung-fish
Turbot
Flat-fish

Sharks

Modern birds

Toothed birds
archaeopteryx

AVES

Dinosauria.

Ophidia & lacertilia
(snakes & lizards)

Pterosauria
(pterodactyls)

Crocodilia

REPTILIA

Marsupialia
(pouched)

Monotremata (egg-laying)

MAMMALIA

Anura (frogs & toads)

Urodela.
(newts etc)

Chelonia
(turtles)

AMPHIBIA

Chondrostei
(sturgeons)

Holostei
(gar-pike)

Teleostei
(most fishes)

Dipnoi
(lung-fish)

BONY FISH

Elasmobranchii etc.
(sharks rays etc)

PISCES

CARTILAGINOUS FISH

Primates

Ungulata

Cetacea

Insectivora

Carnivora

PLACENTAL MAMMALS

Cheiroptera

Rodentia

Edentata

			NO TEMPERATURE REGULATION	WITH TEMPERATURE REGULATION
		NO AMNION	WITH AMNION →	
	NOT TERRESTRIAL	TERRESTRIAL →		
NO BONE	WITH BONE →			
WITH JAWS AND PAIRED LIMBS →				
HEAD, BACKBONE AND ORGANS OF SPECIAL SENSE →				

Milton Keynes UK
Ingram Content Group UK Ltd.
UKHW032321161024
449665UK00001B/6